Science and Technology
in Central and Eastern Europe

Science and Technology in Central and Eastern Europe
The Reform of Higher Education

Edited by
A.D. Tillett
Barry Lesser

NEW YORK AND LONDON

First published 1996 by Garland Publishing, Inc.

This edition published 2013 by Routledge
711 Third Avenue, New York, NY 10017
2 Park Square, Milton Park, Abingdon, Oxfordshire OX14 4RN

First issued in paperback 2016

Routledge is an imprint of the Taylor & Francis Group, an informa business

Library of Congress Cataloging-in-Publication Data

Science and technology in Central and Eastern Europe : the reform of
    higher education / [edited by] A.D. Tillett, Barry Lesser.
            p.   cm. — (Garland reference library of social science ;
    vol. 991. Garland studies in higher education ;   vol. 8.)
        Includes bibliographical references.
    ISBN 0-8153-1768-9 (alk. paper)
        1. Science—Study and teaching (Higher)—Europe, Central.
    2. Science—Study and teaching (Higher)—Europe, Eastern. 3. Tech-
    nology—Study and teaching (Higher)—Europe, Central. 4. Technol-
    ogy—Study and teaching (Higher)—Europe, Eastern.   I. Tillett, A.
    II. Lesser, Barry, 1947–   . III. Series: Garland reference library of
    social science ; v. 991. IV. Series: Garland reference library of social
    science. Garland studies in higher education ; vol. 8.
    Q183.4.E85S38   1996
    507'.1'147—dc20   95-48374
                                                                              CIP

ISBN 13: 978-1-138-98139-3 (pbk)
ISBN 13: 978-0-8153-1768-5 (hbk)

Contents

Series Editor's Preface

Higher education is a multifaceted phenomenon in modern society, combining a variety of institutions and an increasing diversity of students, a range of purposes and functions, and different orientations. The series combines research-based monographs, analyses, and discussions of broader issues and reference books related to all aspects of higher education. It is concerned with policy as well as practice from a global perspective. The series is dedicated to illuminating the reality of higher and postsecondary education in contemporary society.

It is published with the cooperation of the Center for International Higher Education and the Program in Higher Education, Boston College, Chestnut Hill, Massachusetts.

Philip G. Altbach
Boston College

Preface

The present book contains a series of essays on the state of science and technology in Central and Eastern Europe (CEE) and the Former Soviet Union (FSU). The particular topics and countries covered in the book were chosen following discussions between Dr. Tom Eisemon of the Education and Social Policy Department, the World Bank, who was the project's progenitor, and A.D. Tillett and Barry Lesser of Lester Pearson International (LPI), Dalhousie University (Canada).

The studies were commissioned in mid-1992 and draft papers were delivered in the last quarter of the year. This was followed by author revisions and editing, leading to a working document submitted to the World Bank in May of 1993. The document then served as background for a conference sponsored by the German Rectors' Conference in cooperation with the World Bank in Italy in July of 1993.[1]

The state of science and technology in the CEE/FSU is a moving target, and a number of situations may have changed since the first draft was written. Moreover, the latest revisions did not necessarily seek to update the factual base of the papers. But overall, we do not believe that the general situation has changed sufficiently in any of the countries for this to be considered a significant problem.

The views expressed in each paper are, of course, those of the authors. A deliberate effort was made by the editors to try not to alter the meaning in the course of editing the text. Given that the papers were all originally drafted in English while many of the authors do not speak English as their first language, this was not always easy. We apologize to the authors if, in fact, they have been misrepresented in any way through the editing.

The science and technology field in the CEE/FSU is changing quickly (as are most areas in these countries at present), and this book is intended only as a preliminary overview of the situation. As such, it is seen as neither exhaustive nor definitive in any sense of those terms.

We owe a debt of gratitude to Dr. Eisemon for his probing and patient support of this work. In addition, the activity could not have been undertaken without the support of the LPI staff: Pat Rodee and Marion Rout for keeping us within budget and facilitating payments; Nancy Hayter for her work in editing, style, presentation, and overall supervision of the final manuscript; Lesley Partanen and Bernie Misener for inputting and reproducing the first draft; Bernadette Mount and Bernie Misener for work on the final revisions; and Ann Watling for assisting with the literature search. We thank them all.

Note

1.   The proceedings of this conference have been published as *Perspectives on the Reform of Higher Education in Central and Eastern Europe* (Bonn: Dokumente Zur Hochschulreform 90/1994).

Barry Lesser
A.D. Tillett

Contributors

Thomas Owen Eisemon is a Senior Education Specialist in the Education and Social Policy Department of the World Bank, Washington, D.C., U.S.A., and Professor and Director of the Centre for Cognitive and Ethnographic Studies at McGill University, Montreal, Quebec, Canada.

Zbigniew M. Fallenbuchl is Professor Emeritus of Economics and former Dean of Social Science at the University of Windsor, Windsor, Ontario, Canada.

Vladislav Hancil is employed by a Czech insurance company as an evaluator of petrochemical risks. He is a former Vice President of the Czechoslovak Academy of Sciences.

Volker Lenhart is Professor of Education at the University of Heidelberg, Germany.

Barry Lesser is Professor of Economics and Chair of the Department of Economics, Dalhousie University, Halifax, Nova Scotia, Canada, and Associate Director of Lester Pearson International at Dalhousie University.

Elena Z. Mirskaya is Professor of Sociology and Head of the Department of Sociology of Science at the Institute for the History of Science and Technology, Russian Academy of Sciences, Moscow, Russia.

Dmitry Piskunov is the Director of the Analytical Centre on Socio-Economic and Scientific Technological Development of the Russian Academy of Sciences in Moscow, Russia.

Yakov M. Rabkin is Professor of Education, Faculty of Education, at McGill University, Montreal, Quebec, Canada.

Jan Sadlak is a staff member of the Division of Higher Education of UNESCO, Paris, France.

Stephan Stockmann is a planning consultant at the Institute for Development Cooperation, Heidelberg, Germany.

Anthony D. Tillett is Director of the Latin America Regional Office of the International Development Research Centre (Government of Canada), based in Montevideo, Uruguay, and former Executive Director of Lester Pearson International at Dalhousie University, Halifax, Nova Scotia, Canada.

Part One

Introduction

Chapter One
The Context of Change

Barry Lesser and A.D. Tillett

The transformation taking place in the former Soviet Union (FSU) and Central and Eastern Europe (CEE) is affecting all facets of those societies. One area that causes some concern is what is happening to universities and research institutions. More specifically, the larger process of political and economic liberalization now underway in Eastern and Central Europe is, at least in the short term, having a debilitating impact on institutions of science and technology. At the same time, these institutions themselves are experiencing pressures for change in order to adapt to the new political and economic order. An important question is whether they can withstand the negative pressures from the larger process of change long enough to complete their internal transformation so as to become viable institutions in the new economic and political environment.

That is not a trivial question. Institutions of science and technology have a key role to play in the long-term transformation of the economies of Eastern and Central Europe. The Soviet system created a relatively high-level science and technology infrastructure. It was not without problems; indeed, it is an understatement to say that it was a highly imperfect system. But for all of its faults, an infrastructure was developed.[1] This is now in danger of collapsing through a lack of funding, migration of skilled personnel, internal resistance to change, and other such factors. If that happens—that is, the existing system does collapse—the entire process of transformation will falter, not just in the institutions of science and technology but also in the countries as a whole. If, on the other hand, the system maintains itself but fails to adapt appropriately to the new realities, its contribution to the long-term process of change and economic revitalization will be trivial at best, and the larger transformation process will again be threatened. Thus, what happens to the scientific and technological infrastructure in Eastern and Central Europe matters a great deal.

These are the questions that have driven this study of the institutions of science and technology, and, more broadly, higher-education institutions, in the former Soviet hegemony. There is a need to understand the importance of these institutions, their current status, their future prospects, the process of internal revitalization, and their ability to withstand the multitude of pressures they are experiencing. This understanding, in turn, will allow a determination to be made of how best to assist these institutions in the short term in order to ensure that their long-term potential contribution is secured.

This study consists of a series of review papers focusing on some or all of these issues as they relate to different countries of the former Soviet bloc, including Russia. In this first chapter, we try to set the stage for the country reviews that follow by looking briefly at the role of science and technology in the process of economic development and growth. This is followed by brief comments on some of the salient characteristics of Soviet policy and institutions relevant to science and technology, and highlights of the problems confronted and changes occurring in those institutions.

The country reviews that follow this chapter are in turn followed by an examination of the experience and lessons gained from other jurisdictions and a final chapter of summary and conclusions.

Overview of the Role of Science and Technology in the Economy

Economic growth reflects the success which a country has had in mobilizing its resources to achieve technological breakthroughs.[2]

This view of technology and its role in economic growth is a common one, certainly in the context of the historical evolution of the Western industrialized economies. The questions it raises, however, are many, even if we accept the statement at face value; chief among these questions is why some societies seem to have a greater capacity for generating technological change than others.

The answer to this question is far from simple and is still fiercely discussed. But clearly, one of the major factors that explain differences in the rates of invention and innovation among countries in this century is the varying level and quality of those countries' science and technology infrastructures. This is not to suggest that the latter is a sufficient condition for technological progress, nor that the conditions for invention and innovation are even the same—they are not. It is to suggest that a high-quality and adequately sized science and technology infrastructure has become a necessary condition for technological progress and hence, for economic advancement.

Equally important is the link between science and technology on the one hand and the educational system on the other. The educational system is important for at least two reasons:

1.   It is the source of personnel required by a society to undertake the invention, innovation, and diffusion of science and technology.
2.   It is an important focus of direct research that advances the science knowledge and pace of technological advance in a society.

David Landes, in his classic work on technological change, notes that at the commencement of the industrial revolution, the needs of industry confirmed two highly relevant constraints, those of "scarcity of skills and scarcity of venture capital."[3] On the scarcity of skills, Landes notes:

Skills are learned. And the supply of skills to industry is essentially dependent on education. . . . By education we really mean the imparting of four kinds of knowledge, each with its own contribution to make to economic performance: (1) the ability to read, write and calculate; (2) the working skills of the craftsman and mechanic; (3) the engineer's combination of scientific principle and applied training; and (4) high level scientific knowledge, theoretical and applied.[4]

Many studies, too numerous to mention here, have confirmed that engineering and basic and applied science are key areas of knowledge contributing to economic performance. Landes draws our attention to the essential links between skills in the marketplace/workplace and the educational system.

The link from science to technology to the economy is not a simple one. The embedding of much of science within educational and research institutions increases the complexity of the relationships even further. As Douglass North has noted:

The systematic demand for scientific knowledge is a modern phenomenon and is surely related to a growing perception of its usefulness in solving practical problems. A distinctive feature of its institutionalization in universities and research organizations is the recognition of social demands on a broad front. Advances in scientific knowledge must proceed along a wide variety of lines so that our ability to employ developments in one area is not impeded by bottlenecks in another.[5]

This statement underlines the arguments already made regarding the institutionalization of science and technology and the role of scientific knowledge. But North adds another dimension to the relationship between science and economic performance in introducing the notion of "social demands" on science, representing demands that clearly extend beyond the purely economic.

Social demands further complicate the incentive structure in place for the development of science and the transfer of scientific knowledge into usable techniques (technology) and the subsequent adoption (diffusion) of technology. Landes notes that, in spite of the argument that science and technology contribute to economic success, "[t]here is [a] discrepancy between cognitive inputs and economic outputs," that is, the empirical evidence does not necessarily support the hypothesis of the relationship between the two. But the reason for this is precisely the complexity of the relationship as we have been describing it.

The complexity of the role of science and technology vis-a-vis economic progress makes it particularly difficult to "prove" the relationship in any simple, empirical way. Clearly, the private sector and applied-research facilities within corporations have a significant role, as do North's "social demands." But equally clearly, basic research in scientific institutions and applications that flow from such work, as well as the training of new cadres of scientists and engineers, also play an important role.

This very brief overview of the role of science and technology in economic development does not begin to do justice either to the arguments regarding the nature of the relationship nor to the vast amount of existing literature that explores this issue. Nonetheless, several conclusions emerge from this discussion that are relevant to the purpose of this paper:

1.   Science and technology are important to the fostering of long-term economic growth.
2.   Science and technology institutions have an important role to play in realizing the contribution of science and technology to economic growth through their function as educational institutions and research institutions.
3.   Basic science, as well as applied science, contributes to the impact of science and technology and hence, long-term economic prosperity.
4.   Just as important as a specific body of scientific knowledge is the ability of a country or an economy to understand when and how to apply such knowledge to specific problems—North's "social demands." In different words, Robert Reich (1990) has described this point as follows:

Policy makers have failed to understand that a nation's real technological assets are the capacities of its citizens to solve the complex problems of the future—which depend, in turn, on their experience in solving today's and yesterday's.[6]

Such capacity is very much a function of higher education.

An Overview of the Soviet Model
Space does not permit a long discussion of the features of the Soviet model of science and technology. The papers that follow all describe certain aspects of this system in the countries with which they deal. At this point, we simply summarize some of the main features of the system as outlined in these papers and in a few other sources (for example, Parrott[7]):

1.   There was a relatively high, if inefficient, level of integration between institutions of science and technology on the one hand and the planning process on the other.

2.   It follows that the science and technology infrastructure was highly centralized in terms of control and direction—that is, in setting priorities.

3.   Military related research and science dominated the science and technology agenda. Much effort and many state resources were put into the development of a relatively small number of areas of high expertise, while most other areas were underfunded and of low(er) quality.

4.   The system was highly politicized. Researchers advanced in the system as much for their politics as for the quality of their science. At the very least, no researchers could survive, under normal circumstances, if they were judged "politically incorrect"—that is, if they did anything of an overt nature to criticize or undermine the established political order. In the worst cases, such as Romania, the system was absolutely suborned by the political system.

5.   As a consequence of the political factor, scientists of less than top quality often ended up responsible for the system because they were politically acceptable or, in many cases, as part of the overt political reward system.

6.   The scientific community was, to a significant degree, cut off from contact with the international scientific community. At times, the political factor verged on paranoia, leading to immediate suspicion of anyone in regular communication with Western counterparts. Only "politically safe" scientists were allowed to travel to scientific conferences in the West (as a reinforcement for good behavior); these were typically not the persons best able to benefit from such meetings and interactions.

7.   There was a separation of teaching and research. Whereas in the West these functions, at least in terms of basic science, are integrated in university institutions to a high degree, in the Soviet system they were separated. Universities or technical schools performed the teaching and education function, and research academies and institutions carried out research.

8.   For teaching institutions, politics were also an essential part of the curriculum, both in the direct sense of specific and mandatory political courses on Marxist thought, social economics, and the like, and, indirectly, through the prohibition of politically unacceptable content in any other courses. This censorship led to large omissions

or, at the very least, distortions of the material taught to students.

9.   In the research institutions, the scientific agenda was almost entirely driven by the political and central-economic planning mechanisms; that is, there was little or no research autonomy.

10. The system was entirely state funded. Students paid no fees and were in fact supported during their program of study. Equipment, salaries, and so forth, in both the educational and research institutions, were part of the state budget, provided either directly or indirectly via grants from specific ministries or state-owned enterprises. Significantly, the scientific community was relatively well-treated in terms of salary and other benefits—partly a means of ensuring right behavior and partly a reflection of the relative esteem or prestige with which scientists were officially regarded.

11. Because of the highly centralized planning structure used by the Soviets and the predilection to create centers of expertise in certain fields (in an equally centralized fashion), resources tended to be distributed with a high degree of institutional differentiation accompanied by a geographic dispersion throughout the various republics of the FSU and the countries of the CEE. The breakdown of Soviet hegemony and the breakup of the Soviet Union itself left many states with a pocket of expertise in one or a few fields and almost nothing in others. The match with local expertise or economic advantage, moreover, may be nonexistent, as these factors had little or nothing to do with the choice of location or institutional specialization in the first place, let alone the present environment of change. Irrespective of any questions of quality or mission, the Soviet scientific enterprise was the largest in the world; it is now fragmented into the fifteen parts of the FSU and the independent countries of the CEE.

12. The previous point notwithstanding, there was always a tendency to cluster the most important institutions for science research in Russia. This leaves Russia in the strongest position in the new era, but one that is threatened for a variety of other reasons.

13. The economic role of science versus its military importance was never given as much prominence as in the West. Thus, significant links with the industrial structure were (and are) absent in a relative sense; the system was possibly more of a drain than a contributor to the economy.

14. In spite of many of the preceding comments, which individu-

ally and collectively paint a damning picture of the science and research establishment in the FSU and CEE, there were in fact a significant number of qualified persons and much quality science carried out. This is an important point because it is that nucleus of quality and achievement that is the foundation on which arguments to reform the system versus replace it are based. This point should in no way detract, however, from the inescapable conclusion that the range and depth of transformation required is vast.

Problems of Reform

The current status of science and technology, and higher education more generally, in the FSU and CEE demonstrates both challenges and opportunities. In this respect, this sector is a mirror on the larger process of economic and political change transforming these countries. As in the larger national picture, it is the challenges that presently dominate the scene.

In more specific terms, the reform process is currently subject to a number of pressures, which can be roughly divided into two main categories: financial/economic and structural/organizational.

The financial/economic set of pressures arise from a combination of the general restraint on government spending now in place, coupled with (a) the impact of the downturn in overall economic activity together with the relatively high rates of inflation experienced in most cases until at least early 1994; and (b) a greater than proportional decrease in governmental funding for this sector. The latter in part reflects this sector's lower short-run priority relative to other adjustment demands on government, in part a lessening of confidence in the role of indigenous science and technology with respect to economic development, and in part the decrease in military spending, given the previous strong link between scientific research and the military. The lower level of confidence in the indigenous science and technology sector in turn results from a perceived failure by the system to deliver in the past, and the current relative emphasis on the need for imported technology as the quick way to catch up to Western technological levels and standards.

Little can be said about the general restraint on governmental spending, nor is there much to discuss as to the relative priority of competing adjustment demands on a reduced governmental budget; short-term needs will of necessity always win out even though the long-run costs may be compounded in the process. Some comment can be offered, however, on the impact of the current economic climate, the importance of an indigenous science capability, the perceived failures of the past, and the role of foreign technology.

The general economic climate, coupled with the decrease in governmental funding, means that the science and technology sector is confronting a depressed economy while dealing with declining funds and while facing rising costs virtually across the board. Journal subscriptions are not being renewed, book acquisitions are down, equipment costs are rising, and, very importantly, salaries have fallen dramatically in real terms. This is especially important because it is a significant contributing cause to a major migration of scientists to the West, threatening the FSU and CEE with the loss of an important resource. The fact that as of the beginning of 1995 there are signs of economic recovery in many of the CEE countries, while encouraging, does not provide much basis for believing that this problem is solved. Economies may have bottomed out, but they still face a long climb back up; resource constraints together with competing demands for those resources are still very much a fact of life. Moreover, governmental fiscal restraint, which has been an element of the stabilization policy that in part explains the recovery, is not about to be relaxed.

The role of an indigenous science and technology sector was discussed earlier; it is important to appreciate that the perception of past failures is to a large extent the failure of the Soviet centralized planning model, not of the science establishment. It was the product of the system of which it was a part; in other words, it was what the system made it. In certain areas it had its successes—it showed that it was capable of performing when it was allowed or made to do so. This does not alter, however, the perception of failure.

What is also important to appreciate is that, while short-term demands on the system may of necessity have to take precedence, the long-term perspective must not be lost sight of or be neglected entirely. There is a danger of this happening if remedial action to reverse current trends is not taken. We return to this issue in the final chapter as an area where external assistance may have a particularly important role to play.

Finally, the role of foreign technology must be examined both as an argument in its own right and relative to the role of the indigenous sector. What is at issue here is whether foreign technology can or will be the panacea it is thought to be, and whether it is a viable long-run strategy in any case.

In those areas where the FSU/CEE has existing, world-standard pockets of excellence, there is a strong case to be made for maintaining and building on those areas of strength. Overreliance on foreign technology will not achieve this. In addition, sole reliance on foreign (imported) technology will leave countries in the position of playing the role of users rather than producers of technology. When there is no comparative advantage to being a producer of technology, this may be a sound approach. But the size of the science establishment in the FSU/CEE, combined with its pockets of excellence, suggest that there may be a comparative advantage to be taken into account, at least in some areas. Ironically, this fact seems to be better recognized by outsiders than by the countries themselves. Several deals have been made by foreign corporations to support science research establishments in return for claims on the results of the research. While such deals may have their advantages, in certain cases they may be equivalent to the movement of the facility offshore, except for the technicality that it is being maintained at home. In other cases, whole establishments are literally moving offshore, being "bought" by foreign corporations or governments. Foreign interests clearly are perceiving expertise worth having while domestic governments are not. The irony of this is that if all the best leave or are bought up by foreign interests, the judgment that the remaining system is not worth preserving will become a self-fulfilling prophecy. Those not worth keeping, both institutions and individuals, are those that will be left.[8] This has not yet happened, but it could.

The second problem area, that of structural/organizational change, is tied to the financial/economic situation. Declining budgets, rising costs, migration, and falling demand all serve to make structural reform of the system more difficult to accomplish. While it is true that reform may be forced on institutions by adverse economic conditions, the notion that this is a positive development is far too sanguine, not to say simpleminded. The loss of the best people means that one of the most important forces for creative change is lost. Moreover, the type of reform borne of financial exigency may not be the type of reform required to build a new, more humane society.

Financial issues aside, the process of reform is encountering other difficulties:

1.   Resistance to the need to get rid of the ideological and political content of programs because of the vested interests of staff and faculty and because of institutional inertia.

2.   The inability to get rid of untrained or inefficient staff who held their jobs because of their politics in the old order.

3.   The relative lack of experience with institutional self-government after so many years of the centrally directed Soviet system.

4.   The lack of structures to support institutional self-government.

5.   The lack of equipment and supplies, particularly of the latest technological vintage.

6.   The previous separation of teaching and research, and an absence of any strong research ethic or culture in some university departments.

7.   The lack of professional training or skills in areas where the economic transformation of the country is creating a demand.

8.   The breakdown of the previous scientific network as relations are severed among the countries of the former Soviet bloc for reasons of national independence.

9.   Competing claims on the system by ethnic minorities or different ethnic groups, which often fragment the already dwindling resources available to the system.

10.   The fact that, in several countries, both faculty and students in universities were in the vanguard of the movement for change, with the result that they are not necessarily trusted by the political system. More significantly, there can be an excessive belief within the universities in their role as champions of reform, placing them in many cases in the position of competing with government, thereby lessening their support from the political system.

11.   The lack of any well-defined system of standards by which to judge the quality of personnel and the content of the programs.

12.   The need to rationalize the system, including the reallocation of resources both within and without the existing structure.

13. The lack of experience in operating in a private market setting combined with forces encouraging privatization, such as, for example, the introduction of tuition fees.

14. The creation of a number of new, private institutions that are competing with the publicly funded system and, in the short term, drawing away personnel and students, which serves to weaken the public education and research infrastructure, the advantages of competition notwithstanding.

This is neither an exhaustive nor a definitive list of the nonfinancial problems faced by the reform process in institutions of science and technology, and higher education in general. Rather it is illustrative of the range and complexity of the problems faced, problems that will have to be overcome if change is to proceed in the way necessary for the emergence of a revitalized science-and-technology sector, capable of playing the important role it must in the overall process of reform and transformation.

Conclusion

The papers that follow describe in more detail many of the features of the old order and the problems of change noted above. They demonstrate that there is much in common in the problems faced by the various countries, at the same time as they underline the fact that there are no universal solutions: Each country must make its own decisions, and there is ample room for each country to choose its own path. What is common to all the countries of the FSU and CEE, however, is the undeniable role that science and technology, and institutions of science and technology, have to play in the broader process of societal change now underway, through both research and teaching.

As private markets become more developed in these countries, it can be expected that private industry will assume more of a direct role in science and technology activities—that is, in research and development (R&D). That should not be seen, however, as a substitute for the maintenance of a strong and revitalized public role in science and technology. Both are important and each complements the other. Brooks, writing of the United States, makes the following comment:

In principle, I believe the United States still retains the capacity to stay in front of the rest of the industrialized world, but not way in front, if it gives high priority as a society to science, technology, education, and productive investment without sacrificing a reasonable degree of equity among its population. This is not an easy prescription nor is it an impossible one.[9]

Brooks affirms a role for both the private and the public sector; his prescription cannot work without both and without the two working together. Recognizing that everyone cannot be first but that the prescription for success will otherwise be the same, his words have as much or more meaning for the countries of Central and Eastern Europe as for the United States. The challenge for the former, under present circumstances, however, is infinitely greater.

Notes

1. Recall, for example, the optimistic remarks of Zhores Medvedev in the final chapter of his critical book *Soviet Science* (New York: Norton, 1978).
2. L.E. Davis, et al., *American Economic Growth, An Economist's History of the United States* (New York: Harper and Row, 1972).
3. David Landes, *The Unbound Prometheus* (Cambridge: Cambridge University Press, 1966), p. 339.
4. *Ibid.*, pp. 339–40.
5. Douglass North, *Structure and Change in Economic History* (New York: Norton, 1981), p. 172.
6. Robert Reich, *The World of Work* (New York: Random House, 1990).
7. Bruce Parrott, *Politics and Technology in the Soviet Union* (Cambridge, Mass.: MIT Press, 1983).
8. See "Russian Science in Crisis," *Scientific American,* 269 (ii), February 1993, pp. 92–100.
9. Harvey Brooks, "National Science Policy and Technological Innovation," in R. Landau and N. Rosenberg (eds.), *The Positive Sum Strategy, Harnessing Technology for Economic Growth* (Washington: National Academy Press, 1986).

Part Two

Transition: Country Experiences

Chapter Two
Russia
Higher Education and Change

Dmitry Piskunov

The breakdown of the unified educational complex, as a result of the collapse of the USSR, has had a considerable impact on universities and other institutes of higher education. Although higher education was administered by different republican and all-union or national government ministries, all fundamental decisions (such as creating new institutes, specifying the nomenclatures of specialties, defining the quantity of admitted and graduated specialists, elaborating educational programs and curricula, preparing and issuing textbooks and professional magazines, guaranteeing graduate employment, and so forth) were taken and maintained on the all-union level.

Soviet higher-educational institutions in Russia, as well as the other republics, functioned as components of this complex. Following its disintegration, the Russian government has not been able to rapidly reform or reconstitute a system to manage higher education. Higher education is no longer a primary object of state or public concern. It is quite probable that, if the current tendencies persist, the internal educational crisis will acquire avalanchelike proportions.

General State of Higher Education

Financing, Material, and Technological Support Russia's higher and professional education system is one of the most developed in the world. At present, it is made up of 534 institutes of higher education, which accommodate 2.8 million students. The basic indicators can be found in Table 2.1.

The state continues to play the major role in providing financial assistance to educational institutions. For the last several years,

Table 2.1 Basic Indicators of the Development of the System of Higher Education in the Russian Federation, 1980–1991

	1980	1985	1989	1990	1991
Number of institutes of higher education	494	502	512	514	519
Number of students*	3.046	2.996	2.795	2.824	2.763
Enrollment of students*	613	634	—	583	566
Competition (number of those willing to enter (per 100 places)	188	166	189	194	204
Number of graduates*	460	477	433	401	407
Number of teaching staff*	204	205	216	220	224
professors, doctors of science	9.6	9.8	12.9	13.7	14.2
associate professors, candidate of sciences	92.4	103.0	114.6	115.3	115.1
Graduates per 10,000 population (persons)	33	33	29	27	—
Specialists with higher education per 1,000 persons capable of working	95	110	125	—	—

*In thousands of persons.

the institutes of higher education have been allotted 1.6 to 1.8 percent of the annual budget. Their expenses have increased slightly in the last decade (see Table 2.2). Current expenses (rent, heat, student stipends, and so forth) constitute three-quarters of the budget, and capital expenditures (equipment purchases and other investments) have not exceeded 9 percent. Wages for scientific and teaching staff have increased slowly when compared with those in industry.

Education (including higher education) has repeatedly caused heated debates in the Russian Supreme Soviet and the media. Statesmen, representing various echelons of political power, have vowed that the new Russian leadership will improve and radically reform the system of higher education.

Education was declared a major state priority by the first decree of the president of the Russian Federation, Mr. Boris Yeltsin, in July of 1991. The decree envisaged significant investments in higher education, permanent salary indexation, and maintenance of academic wage levels above the national average. By 1992 it was clear that these promises could not be kept.

Efforts to improve the situation have not been fruitful. The government decided, for example, to allot 4.4 million rubles in 1991 (versus 1.5 million in 1990) to the institutes of higher education supervised by the Committee on Higher Education of the Russian Federation. In the meantime, according to expert evaluations, prices had risen by almost 800 percent, with no qualitative improvement.[1]

Table 2.2 Indicators of the Maintenance Costs of the Institutes of Higher Education Covered by the State Budget of the Russian Federation

	1980	1985	1989	1990
Total maintenance costs (billions of rubles)	1.5	1.6	2.1	2.4
Cost of one student (rubles per year)	1,076	1,156	1,463	1,861
Professors' average monthly salary (rubles)	450	450	550	550

Although higher education's share in the state budget grew from 1.9 to 2.7 percent between 1990 and 1992, that sum could not compensate for inflationary losses. While nominal budget allocations for higher education have continued to increase since 1992, continuing high rates of inflation in Russia have meant a continuation of this problem; in real terms, higher education has experienced a significant decline in funding. Partly for this reason the last several years have seen a drop in the number of new buildings for higher education (a decline of 32 percent between 1980 and 1989). In many cases, institutes of higher education are renting buildings and students are studying in two shifts.

Replacement of training equipment in Russian institutes of higher education has always been a slow process (seventeen years on the average); it is even slower today. According to "Higher Education in Russia: Conditions and Program of Development," a report prepared by the Committee on Higher Education for the president, institutes of higher education obtained equipment worth 55.2 million rubles (instead of the planned 71.1 million) in 1991, with only 22 million rubles from the state budget (it should have been 32 million rubles). In 1992, the institutes received only 14 percent of the capital investment funding they required. Equipment imports for educational needs stopped completely. In 1992 not a single contract was signed with a foreign firm, because of the absence of hard currency.

The institutes of higher education used to receive additional financial resources from their direct economic links with ministries and large-scale industrial enterprises (primarily in the defense sector). A large number of laboratory buildings and complete institutes were constructed and equipped by those sources. That support is disappearing under the weight of the harsh financial crisis, thus leading to the problem of preserving employment of specialists.

Training of Specialists Specialists in Russia are educated in a number of ways. The overwhelming majority of institutes of higher education have evening and correspondence departments that trained, in the past, 49 percent of enrolled students. During the last five years, enrollment in correspondence branches declined by 20 percent and evening branches by 50 percent.

For the first time since the Second World War, decreases in the number of enrolled students in the institutes of higher education were recorded, and the decline has affected the evening, correspondence, and daytime branches of the institutes. The total number of students and graduates has declined. In 1991 there were almost 7 percent fewer students than in 1985. These declines will affect how departments catering to full-time students function in the future.

Institutes of higher education and specialists are distributed quite unevenly across Russian territory. Russian institutes are located in 122 towns, with just 15 locations accounting for 49 percent of the total number of institutions (Table 2.3). In the former USSR these concentrations could be regulated, but the nature of the problem is drastically different when Russia has to deal with institutes and students in different, now independent, states. As an illustration, the number of students from these newly independent states has shown a significant drop. In 1986, some 30 percent of the first-year students at Moscow Institute of Electronic Technology came from republics outside Russia; in 1992 their share constituted less than 10 percent.

With the exception of Russia and probably Ukraine, not a single republic of the former USSR has adequate numbers of specialists, particularly as far as high technology is concerned. With the collapse of the traditional links, most newly independent states will inevitably face great difficulties in trying to found their own national systems of higher education. At the same time, in a number of newly independent republics the teaching of Russian is being rapidly curtailed, and the establishments of higher education are being

Table 2.3 Main Centers of Higher Education in Russia

City	Number of Institutes of Higher Education
Volgograd	10
Voronezh	13
Ekaterinburg	14
Kazan	13
Krasnoyarsk	12
Moscow	74
Nizhniy Novgorod	13
Novosibirsk	15
Omsk	14
Orenburg	9
Rostov-on-Don	13
Samara	13
St. Petersburg	42
Saratov	12
Khabarovsk	11

monopolized by indigenous languages. A reduction in teaching standards is quite inevitable when one takes the absence of textbooks and scientific and methodological literature into consideration. The outburst of nationalistic sentiment is extremely dangerous in the sphere of education. The case of Estonia shows the extent to which the position of the nonnatives can be jeopardized: The legislature, rigidly restricting the right of citizenship and introducing Estonian as the only language of education, has deprived a substantial part of the population access to educational institutions.[2]

A similar trend is to be found in the other newly independent states. If national language becomes the test, the children of those twenty-five million Russians and, in total, the sixty million former Soviet citizens who do not belong to the major nationalities of the states in which they live will be excluded from general education. These persons will be forced to change their residence, thus enlarging the number of refugees in Russia. In Russia, competition is the basis of selection. School graduates who wish to enter institutes of higher education have to take obligatory entrance examinations. The last two to three years have seen a slight increase in the number of those wishing to enter such institutes. However, interest differs by subject: Competition has risen for economics (from an average of 2.2 persons per place in 1985 to 3.5 in 1991) and law (from 3.4 in 1985 to 4.8 in 1990), while competition for the engineering professions has fallen to an average of 1.7 to 1.8 persons per place.

Do these trends reflect specific situations or real motivational changes among youth? A number of experts argue that higher education is becoming less attractive for young people, and the evidence of a growing dropout rate confirms this trend.[3]

The content of teaching is also changing dramatically. In 1991 more students enrolled in natural sciences, the humanities, and economics than in engineering (see Table 2.4). Within engineering, traditional specialties still predominate, while the most technologically advanced fields have declined as the demand from military-industrial enterprises has collapsed and has not been replaced by civilian demand. Given the lack of investment and the inefficient role of private interests, the decline will persist. As a result, Russia will be kept outside the world of high technology.

The present-day system of retraining and reskilling lags behind demand. According to some forecasts, up to thirty million people were to have lost their jobs between 1993 and 1995, and it will be increasingly difficult to retrain and redistribute workers in new jobs.[4] The problem is not only quantitative but also qualitative: The system as a whole is not adjusting to the demands of the new economic environment. Even though two million people graduated from retraining and professional courses between 1989 and 1990, that met only 25 percent of the needs of the basic specialties oriented to traditional technologies, and 1 percent of the needs of advanced technology fields. The gap is still growing.

Higher Education: Scientific and Teaching Personnel The institutes of higher education of the Russian Federation employ about 525 thousand persons; teaching staff consist of about 224 thousand, of whom over fourteen thousand are professors and doctors of science and almost 116 thousand are doctoral candidates. The institutes employ almost half of all the doctors and doctoral candidates working in the Russian Federation.

Although the personnel of the institutes of higher education continue to be among the most stable and teaching and scientific work still enjoys high prestige, there are growing signs of change. Since 1990, a growing number of skilled people have quit their jobs. More than 10 percent of the teaching staff without scientific degrees have retired since 1985. In 1992 more than three thousand highly skilled teachers left Moscow institutes. The thirty-year-old teacher

Table 2.4 Changes of Enrollment Structure in Regard to Specialties (in percent)

	1980	1985	1989	1990	1991
Enrollment as a whole	100	100	100	100	100
Natural sciences and humanities	31.2	32.4	35.1	36.6	38.5
Health care	6.7	6.3	7.4	7.5	7.7
Culture and arts	2.9	3.0	3.0	3.1	3.1
Economics	8.6	8.7	8.6	8.8	9.6
Engineering	49.6	48.6	45.0	43.1	40.2
the most up-to-date branches	9.3	9.0	11.3	10.9	10.2
Agriculture and forestry	1.0	1.0	0.9	0.9	0.9

with a scientific degree is becoming a rarity, and the average age of teachers is increasing. The average age of a doctor of science is 58.2 years, near pension age. According to the most recent information provided to the Conference of Rectors of Russia's institutes of higher education (November 1992), 46 percent of engineering science teachers (most of whom held the degree of "candidate of sciences," and were below the age of forty) expressed a desire to quit their job (compared with 18 percent in 1989).

Poor salaries are a major reason for the departure of teaching and scientific personnel. In 1990, their average monthly salary was 236 rubles, 79.7 percent of the average industrial wage; in 1985 it had been 90.9 percent. Moreover, the gap has increased further since 1990. At the beginning of 1991, a professor's salary was 1.8 times the average industrial salary. By the autumn of 1992, scientists earned 4,500 to 5,000 rubles per month on average, while an industrial worker received 5,870 rubles. Most scientists earn only one salary, which now is barely enough to cover essential living expenses. As price liberalization continues, 75 to 85 percent of the personnel in institutes of higher education, and their families, are being faced with declining living standards.

A widespread dissatisfaction with work is another powerful factor convincing many to retire. The 1992 Committee on Higher Education report shows that only 10 percent of the [candidates of sciences] and 20 percent of the doctors of sciences believe they have realized their potential. One out of five specialists would emigrate if a satisfactory opportunity arose. According to press reports, the estimated losses caused by specialist emigration exceeded seventy million rubles in 1990 and were greater than the combined total of investments and aid from abroad. It is estimated that some 200 to

250 thousand specialists with advanced degrees are ready to emigrate from Russia every year; at least 20 to 30 thousand belong to the stratum of most qualified specialists and their departure could significantly weaken the country.

There are about sixty-eight thousand postgraduate students in Russia. Seventeen thousand complete their research work annually. Three quarters of the scientists receive their education in postgraduate courses at the institutes of higher education. In the last twenty years, the number of graduates has not changed, although the numbers of teaching and research personnel have increased by 65 percent. As a result, the number of postgraduate students per 1,000 personnel in higher education dropped from 107 in 1970 to 66 in 1989. During the most recent five-year period, the overall number of postgraduate students has declined by 5 percent, and the number of those completing their research work has decreased by 11.7 percent. If these trends persist, the most highly skilled teaching and research staff, such as the doctors of sciences (the highest scientific degree in Russia), could be reduced by 2 to 3 percent a year, resulting in the destruction of the Russian intellectual nucleus.

The Development of Scientific Research in the Sphere of Higher Education Scientific research in Russia has traditionally been divided among three areas: (1) study centers of the Academy of Sciences; (2) research and design establishments directly funded by industrial and economic sectors; and (3) research departments of the institutes of higher education. These three have functioned independently of each other, responsible only to their state sponsors.

Institutions of higher education used to get around 8 to 10 percent of total science funding even though they employed some 35 percent of research personnel, or almost 50 percent of all the doctors and candidates in the sciences. The Academy of Sciences received about 10 percent, with the lion's share (up to 85 percent) of allocations funding the numerous sector study centers.

Scientific equipment per researcher in 1989 was 27.5 thousand rubles for the first group, twenty thousand rubles for the second, and only five thousand rubles for the third. This division of labor hinders scientific potential, undermines the organic cohesion of science and teaching, and impairs the capacity of the teaching staff in the preparation of specialists.

The research institutes do not always take part in priority scientific programs. In 1991 their share in these programs did not exceed 1.1 million rubles out of a total of 39.6 million, or 4 percent. They received a higher proportion of program funding in high-energy physics, high-temperature conductivity, and bioengineering. Financial redistribution used to occur because of informal links between institutes of higher education, the academy's study centers, and economic branches. Many institutes cooperated on a contract basis with ministries or enterprises, especially in the defense industry. Now these resources are extremely limited.

Different styles of cooperation, responding to the growing market for scientific and technological products, are now emerging. For example, institutes of higher education are creating scientific-technological cooperatives, innovation and consulting centers, and so forth. A first such cooperative was established in 1986, and by 1991 there were more than six hundred. In some cases, however, newly established organizations have severed all connections with the formal scientific research structure and utilize the acquired property for primitive profiteering.

A comprehensive analysis of institutes of higher education that was carried out in 1990 shows a more than threefold drop in the number of promising complex research themes. Copyright licenses and patent applications were also down. At the same time, short-term studies grew drastically. Only 14.1 percent of the new designs of 1991, according to Committee on Higher Education data, corresponded to world market requirements. About 70 percent of the R&D projects completed during the year contained no inventions.

The low effectiveness of scientific research carried out within establishments of higher education is strongly conditioned by the unresponsiveness of the existing economic system. The annual rate of introduction of full-fledged innovations designed by scientists from institutes of higher education does not exceed 50 percent. The State Committee on Statistics of the Russian Federation calculates that the inability to introduce new inventions in 1990 resulted in losses— calculated on the basis of average enterprise efficiency (from fifty up to two hundred thousand per unit)—amounting to a total of more than four million rubles.

State Policy in the Sphere of Higher Education

Economic growth, together with science and social development, began to suffer problems in the early 1970s; the roots of that were linked to deficiencies in education. A response was attempted with "Basic Directions of the Restructuring of Higher and Special Secondary Education," approved in 1986/87. The government implemented this program with a combination of policies that included management decentralization, an emphasis on individual responsibility, curricula reform, development of economic self-sufficiency for higher education, and political autonomy for the institutions.

The basic guidelines of this program remain valid today. However, both the law and policy assumed a major role for the state. It was assumed, in particular, that the state would provide permanent financial resources and pay, under a contract system, for training. Economic branches or sectors and their enterprises were expected to partially indemnify institutes of higher education for training of personnel (no less than 3,000 rubles per person) as a stimulus to improve the quality of training. The reform failed, as it relied on central management. It was to be expected that attempts to reform higher education from the top would give no results.

A new program was introduced in 1991—the "State Program of the Development of Higher Education in the Russian Federation"—which defines a strategy for the development of higher education in the Russian Federation up to the year 2005. The primary purpose of the new reform is to provide strategic guidelines and steps to ensure the survival of higher education under the new and austere financial conditions. The principal features of this new plan are as follows:

1.   Strengthening links between the institutes of higher education; setting up a managerial structure to solve problems of higher education systematically; and utilizing limited financial resources with reference to state priorities and urgent social needs.

2.   Singling out among the institutes of higher education a group for primary state support; apportioning financial and other resources with regard to quality and teaching and research potential, using the most important problems of the Russian Federation (republics and major economic regions) as criteria.

3. Reorienting institutes of higher education toward regional and municipal development and bringing their work gradually in line with local financial backing;

4. Promoting ways to encourage the financial self-sufficiency of institutes of higher education, including the preparation of specialists on a contract basis in which expenses are covered completely by the respective state, enterprise, firm, and individual, and abolishing all restrictions that prevent institutes of higher education from making independent use of their research facilities and intellectual property.

5. Introducing a network of institutes of higher education independent of the state, to receive support through accreditation and licensing of their education programs together with various taxation and credit privileges for both the institutes and their students.

These strategic guidelines are being further developed by the Russian government and the Ministry of Science, Higher Education and Technology Policy. The key organizational principle is institutional autonomy and self-management. However, this principle is threatened by the current financial crises; without state support, institutes of higher education cannot restore their viability.

Governmental policy is aimed mainly at establishing a multi-level educational system and introducing bachelor's and master's degrees. These changes are being made by forming new types of educational establishments and reorienting the existing ones. A number of institutes have changed their status to universities and academies. For example, the Cherkes Technological Institute, the Moscow Institute of Invalids, and the Russian State Social Institute were opened in 1991. The Moscow Historical Archives Institute was transformed into the Russian University of the Humanities.

For the first time in the history of Russian higher education, governmental control is not paramount. The Committee on Higher Education is only a department of the Ministry of Science, Higher Education, and Technology Policy, so it is not possible for it to develop policy relations on an equal footing with other governmental bodies. As the Supreme Soviet of the Russian Federation has not yet passed a higher education law of importance, the state has given up most regulation in its withdrawal from educational control.

The principle of state noninterference is the distinguishing feature of present higher-education policy. Such educational reforms, even in stable times, would be a severe test for society and would require, at least, purposeful financial backing, public support, and flexible administration. These conditions are not present today. Institutions of higher education are being required to change their approach and develop a market for intellectual labor at a time when buyers are absent.

Institutes of Higher Education and the Market Economy: Examples

In order to examine these changes more fully, the behavior of thirty institutes of higher education were examined. As might be expected, the institutes differ in their responses to the new economic situation.

The institutes of economics are in a relatively comfortable position, almost experiencing a boom. They are attracting heightened attention from ministerial authorities and large-scale commercial organizations. Methodological and material assistance is also coming from foreign partners. These establishments of higher education are creating new educational units and a widespread network of commercial educational activities.

The Academy of Markets and Management, created in December of 1991 on the initiative of the Moscow Academy of Economics, the State Academy of Management, the Finance Academy, and the Youth Problems Institute, was one of the first educational units to receive a license for the training of personnel. It has received material support from industrial enterprises, commercial banks, and public organizations. A Spanish university is also a founder, and the academy is cooperating with Boston University (U.S.A.), Kerber College (Germany), and a number of other foreign scientific and educational centers in preparing the Russian version of an international MBA (master of business administration) program.

Various conflicts have emerged between the wholesale adaptation of Western educational programs to Russia and the views of Russian scientists regarding their role in the transition of their country to a modern market economy. A solution to these differences

Table 2.5 Basic Indicators of the Functioning of a Number of Institutes of Higher Education in
Moscow and St. Petersburg for the 1991/1992 Academic Year

Name	Number of Teaching and Research Personnel	Number of Graduates	Number of those Finishing Post-graduate Courses	Volume of Research (millions of rubles)
Moscow University	8,817	3,344	—	196
St. Petersburg University	4,143	2,748	380	59.6
Moscow Institute of Physics and Technology	576	753	139	13.3
Moscow Institute of Electronic Technology	1,353	672	63	41.9
Moscow Power Engineering Institute	3,080	2,177	201	74.2
St. Petersburg Institute of Electrotechnics	1,549	1,102	350	35.9
State Academy of Oil and Gas	1,735	1,122	143	46.4
Moscow Mining Institute	496	900	89	53.6

will require, in most economic institutes, a painful process of qualitative restructuring.

The universities, institutes of engineering, and other educational institutions are in a different position. Historically, two major centers have dominated: Moscow and St. Petersburg. These two cities account for some 24 percent of all the institutes of higher education of the federation, with most of them having national or all-Russian importance as well as being located in the centers of the most intensive social reforms. Table 2.5 describes some of these institutions. Their importance gives them weight in the development of the system of higher education as a whole. In spite of a change for the worse in their material and financial situation, these institutes have managed to maintain their authority and stability up to the present day.

Tables 2.6 and 2.7 give an idea of the dynamics of the graduation of students and postgraduates for Moscow and St. Petersburg. The significant drop of graduates between 1989 and 1991 reflects the introduction and subsequent revocation of the draft. Today the situation is returning to normal.

Table 2.6 Graduation Indicators in Institutes of Higher Education in Moscow and St. Petersburg, 1988–1991

Name	1988	1989	1990	1991
Moscow University	4,435	3,549	3,344	—
St. Petersburg University	3,043	2,680	2,670	2,748
Moscow Institute of Physics and Technology	739	765	764	753
Moscow Institute of Electronic Technology	756	729	681	672
Moscow Power Engineering Institute	3,627	3,542	2,422	2,177
St. Petersburg Institute of Electrotechnics	1,718	1,772	1,169	1,102
State Academy of Oil and Gas	1,460	1,115	917	1,122
Moscow Mining Institute	906	695	716	817
Total Number* for the Institutes Subordinated to the Committee on Higher Education	235.6	227.0	209.2	207.9

Source: Individual institutions and Committee on Higher Education.
*In thousands of persons

Research work takes place on a relatively stable basis (see Table 2.8). The high quality and scientific potential of this group is confirmed by the substantial portion of fundamental research activities (20 to 80 percent). However, it appears that even in this relatively well-to-do group of institutes, the internal sources of self-development are coming to an end. Now they have to enter the market unprotected. Most of them earn little or nothing from student training or enterprise research contracts.

Table 2.7 Graduation Indicators for Postgraduates in Moscow and St. Petersburg

Name	1988	1989	1990	1991
St. Petersburg University*	414	406	428	380
Moscow Institute of Physics and Technology	151	156	169	139
Moscow Institute of Electronic Technology	46	50	41	63
Moscow Power Engineering Institute	203	214	255	201
St. Petersburg Institute of Electrotechnics*	428	427	422	391
State Academy of Oil and Gas	111	118	117	143
Moscow Mining Institute	100	102	111	102
Total Number for the Institutes Subordinated to the Committee on Higher Education	6,743	6,682	6,392	6,225

Source: Individual institutions and Committee on Higher Education.
*Total number of postgraduates in the institute

Table 2.8 Structural Characteristics of Research Work in Institutes of Higher Education in Moscow and St. Petersburg

Name	Value of Research (Millions of rubles)			Structure of Research (percent)		
	Total	Budget	Contract	FI	AR	EW
Moscow University	196	166	30	80	20	—
St. Petersburg University	59.6	50.6	9.0	80	20	—
Moscow Institute of Physics and Technology	13.3	11.8	1.5	20	70	10
Moscow Institute of Electronic Technology	32.2	6.5	25.7	25	52	23
Moscow Power Engineering Institute	74.2	32.2	42	25	65	10
St. Petersburg Institute of Electrotechnics	35.9	30.6	5.3	10	80	10
State Academy of Oil and Gas	46.4	5.2	41.2	10	80	10
Total for the Institutes Subordinated to the Committee on Higher Education (Millions of rubles)	5.841	1.669	4.172	21.9	54.3	23.8

FI—Fundamental Investigations
AR—Applied Research
EW—Experimental Work

Moreover, it is becoming increasingly difficult to find jobs for graduates. In 1992, despite all the prestige attached to them, these institutes of higher education failed to place 15 to 20 percent of their graduates. In the near future, that figure could increase to 60 percent. Over 50 percent of those registered as unemployed in Russia have higher or special secondary education. The study centers of the Academy of Sciences have reduced their demand for specialists to 30 percent of the 1988 level (see Table 2.9). These centers could not find work for 110 graduates of institutions of higher and special secondary education in 1988, increasing to 536 in 1991.

The institutes of higher education in question face considerable pressure. They are trying to adjust their teaching and research consistent with financial support. At the same time they must market their educational services and try to establish links to scientific-technological enterprises. They are trying simultaneously to raise additional financing, raise the level of teaching, and attract young people.

As part of this process, the institutes have gained experience in organizing enriched or "higher colleges" for the most talented and a

Table 2.9 Job Placement of Graduates of Institutions of Higher and Special Secondary Education in Study Centers of the Russian Academy of Sciences

	1988	1989	1990	1991	1992
Number of requests	3,348	3,355	2,906	2,048	1,000
Number denied	110	230	502	536	—

more flexible faculty able to teach the advanced programs that do not necessarily correspond to the orientation of a given institute. One such college is attached to Moscow University. Its aim is to increase the level of fundamental knowledge in the field of materials studies. A higher college of management has been opened at the Moscow Institute of Electronic Technology. In 1992 its annual income was 940 thousand rubles, which was almost equal to the sum of all of the tuition fees paid to the courses of the basic faculties (1,102 thousand rubles).

A contract system of retraining and professional improvement signed by higher educational institutes with enterprises and organizations is becoming a source of self-financing for higher education, together with the teaching of foreign nationals on a commercial basis ($1,000–$4,000 per person—much less expensive than in many other countries).

The year 1992 saw the first attempt to introduce tuition fees for Russian citizens. At Moscow University it was supposed to apply to 15 percent of total enrollment. The cost of training to be indemnified by the student amounts to almost 100,000 rubles. In these courses, two hundred students were enrolled (less than 5 percent of the total) for a narrow range of specialties—law, economics, foreign languages—and the university received about two hundred million rubles or 0.8 percent of the university's annual budget (2.5 million rubles). Moscow University is the best educational establishment in Russia, but the example shows that this source of financing is not yet solidly based. Meanwhile, the government plans that tuition fees will cover 15 percent of the expenses of institutions of higher education; the idea appears utopian for the near future.

Another promising avenue is to link research and teaching. With the active participation of the Academy of Sciences, many of the above-mentioned institutes of higher education have initiated a series of centers for both studies and teaching, concentrating on the most up-to-date areas of research and technology.[5] In 1989, for example, the State University of Moscow (as the head organization) and Moscow-based institutes of higher education in the fields of engineering and physics, power engineering, steel and alloys, chemistry, and technology, created a study and teaching center called Kadry. The center prepares research personnel to work at institutes of higher education and Academy of Sciences study centers in high-temperature superconductivity, as well as the development and application of new technologies. This training and professional skill improvement is contracted with concerned ministries and enterprises. Another example is the Fundamental Research on Software, Algorithms, and Physical Problems, a study and teaching center that enjoys the facilities of the Moscow Institute of Physics and Technology. The Institute of Automation of Project Design of the Russian Academy of Sciences as well as a number of scientific and production associations are also taking part. A study and teaching center for the fundamental problems of microelectronics at the Moscow Institute of Electronic Technology has been established. It involves chairs and research units of the institute, a laboratory of the Institute of General Physics, and other study centers of the Russian Academy of Sciences. This center is engaged in research, personnel training, and specialist qualifications, as well as sharing cooperative links with design and production organizations in the field of electronics technology and materials.

In the mining sector, the Moscow Mining Institute and the Institute of Problems of Complex Mineral Extraction of the Russian Academy of Sciences have created a study and teaching center for fundamental and applied research. High-quality teaching is achieved by the active participation of teaching staff, the involvement of postgraduates and students in the more important studies on mining and mining technology, and the direct involvement of scientists of the Academy of Sciences in the teaching program. A similar organization has been established for the oil and gas area with the aim of speeding up the scientific and technological development of the industry and expanding the fundamental and future studies of oil and gas extraction.

Despite such innovations as those just described, scientific and engineering personnel, often the best trained, are continuing to leave institutes at an accelerated rate. In the last three years, more than 20 percent of the teaching and research personnel of Moscow University have left, and about 60 percent of professors and scientists are compelled to take second positions to at least partially compensate for low wages. So there is a real concern about maintenance and replenishment of the teaching and research staff.

Radical economic changes have stimulated fundamental shifts in the sphere of applied research and development, including the rise of innovation and consulting services and expert assessments. Scientific cooperation could play an essential role in fostering science and technology parks. The first proposals for these were made in the early 1980s, but the necessary legal and economic preconditions were not yet present. Now these parks are being established.[6]

Science and technology parks have been put into operation at the Moscow Power Engineering Institute and the Moscow Institute of Electronic Technology, as well as at institutes of higher education in St. Petersburg and other Russian cities. In 1992 Moscow University formed the joint-stock company "Science Park MGU Ltd." (MGU is the Russian abbreviation for Moscow State University). Among the cofounders were Moscow State University, RIKO (a high-risk investment company), NISON (an association with foreign countries in science, technology, and education), and ALKOR (a research and production association). Over 60 percent of the shares are owned by Moscow University. The science park is expected to become a feeding ground for high-technology design and industrial production firms; it intends to launch large-scale projects in biotechnology, laser technology, ecology monitoring, and telecommunications.

These new teaching and research organizations are expected to compensate for budgetary cuts as well as the collapse of long-established economic links by attracting foreign and national private investment and associating institutes of higher education with productive activity. These arrangements have to be flexible. For example, in 1992, the Academy of Oil and Gas obtained permission to sell ten thousand tons of oil as payment for personnel training and research. The academy hopes to realize about 110 million rubles. A higher education institute that trades in oil is rare, but, in today's conditions, institutions are compelled to undertake new tasks in order to survive. Now the academy, together with a Finnish firm and a group of national partners, is hoping to create a drilling materials plant which, if successful, will make the academy an owner with capital.

Other institutes of higher education are considering the possibility of creating enterprises. For example, there are plans to fabricate microelectronic parts at the Technology Center of the Moscow Institute of Electronics, which has good working relationships with universities in Europe and the U.S.A. However, there is a growing understanding that survival depends less on individual strategies and more on collective self-organization. Institutions of higher education are setting up various associations on a corporate basis (associations of universities and engineering, medical, and other institutions). An International Academy of Sciences of Higher Education has been created that unites the scientists of Russian institutes of higher education and centers of the Independent State Community. In November of 1992, the Union of Rectors of Higher Educational Institutes of Russia was formed. The first congress of this union supported the idea of creation of the State Committee of Higher Education and agreed to concede the rights of hiring consultants and experts when developing state policy to this social organization. This idea was supported by the president of the Russian Federation.

This type of reciprocity or collective action could begin an effective and democratic system for the management of higher education in Russia.

General Conclusions

Several conclusions can be drawn from this discussion, the first of which is that the system of higher education in Russia is one of the most powerful in the world; its intellectual potential is enormous. It includes a number of teaching and research centers of world importance and schools of scientific thought with worldwide fame. In many branches of science its personnel are among the most highly qualified, and circumstances are propitious for both the preparation of a wide variety of specialists and realization of large-scale interdisciplinary studies.

This combination of leading scientists and young, talented researchers and engineers provides an opportune basis on which to maintain the complete scientific-technological cycle, from concepts to implementation models to innovation.

Despite the deterioration of economic conditions, the institutions of higher education can preserve their relative stability. The majority of institutions are creating the necessary organizational changes to generate various forms of self-financing and to link, economically, with consumers and industry.

A loss of demand for qualified specialists and scientific production, as well as the absence of effective governmental support, has stopped the reforms of higher education, however, reflecting the loss of status as a strategic priority for economic and social development.

Complete autonomy and self-sufficiency for these institutions at a time of grave crisis is premature. The contrary view is based on the erroneous thesis of "inexpedient conservation of the redundant scientific potential in Russia" and gives rise to the disastrous policy of "liquidation of supposedly surplus science and education."

It is more and more evident that the future of science and higher education completely depends on the state and its policies.

A most important policy guideline is the Federation Agreement, which provides for science and education development to be devolved to the Federation State Bodies and the subjects of the Russian Federation (republics within Russia, the provinces, some cities, and so forth). The legal and financial basis for this is almost completely absent, however, and that is why any decision taken at the federal level alone could be explosive.

Education was among the first social institutions to begin a profound reconstruction based on the principles of democratization and self-development. Many institutes of higher education have reorganized in order to allow self-financing, technology transfer, consumer links, and applied science on the basis of commercial principles.

A readiness to work under market conditions, however, does not guarantee practical success. The implementation of radical reforms is at stake because of the general social crisis. The indifference of the economic system and social and political institutions to the criteria of competence, education, and professional skill will continue to be responsible for the difficult material conditions of institutes of higher education in the near future.

An analysis of the current economic trends shows that the creation of a knowledge market is a remote option. The demand for specialists as well as scientific and technological output is practically absent. At the same time, further cuts in the financial resources for education are inevitable, given:

1.   The fact that the state program of anticrisis measures (1993) plans to raise expenditures for higher education from 2.7 percent to only 3.34 percent of total spending. That is not sufficient, given current rates of inflation, to keep real support at current levels.

2.   The fact that the possibilities of contracting with various enterprises and their associations are being exhausted. Many state-owned enterprises are stagnant, and others can scarcely manage to cover current expenses. According to the State Committee on Statistics, only 12 percent (1.2 million rubles) of the long-term credits given to enterprises from January to August of 1992 were used to expand, reconstruct, or modernize production; 88 percent was used for current expenses. Education and science can hardly prosper under these conditions.

3.   That the growing private sector relies mostly on financial, commercial, and mediatory operations for its prosperity. In other words, it does not rely on education and science. Therefore, private investors and foundations do not offer higher education a significant source of financial potential. This is one reason why the system of higher education cannot overcome its present-day crisis without heavy losses—continued financial deterioration, demoralization of personnel, reductions in the number of students, and so forth. The most urgent problem facing higher education is its survival and size, rather than its management and reform.

There are ambitions for Russian institutions of higher education to enter the world market of education services, as well as science and technological-based production, as part of the global scientific system. However, the specific nature of institutes of higher education, characterized primarily by their teaching and research programs, excludes the mechanical application of organizational arrangements used in purely research cooperation between Russian and foreign partners. The latter mostly involve individual researchers and study groups going abroad.

As far as institutes of higher education are concerned, the general approach of relying on international partners will not do. It is harmful if leading professors of institutes go to work abroad for a long while, or if an institute as a whole changes the nature of its research in response to the interests of a foreign partner. Other options should be considered. Among the most promising are:

1.   Participation of Russian institutes of higher education in projects of UNESCO, the International Association of Universities, the Conference of Rectors of Europe, and similar international bodies.

2.   Preparing personnel and providing research for economic, scientific, and technological projects, in accordance with the United Nations, the World Bank, the International Monetary Fund (IMF), or by agreements with other states.

3.   Involvement of Russia's institutes of higher education in expert evaluations of projects undertaken on its territory in accordance with the requirements of joint ventures, and foreign firms' taking part in project development by means of forming special educational programs and organizing joint educational bodies attached to institutes of higher education.

Some elements of such cooperation exist today but only among few institutes of higher education, and without being associated with a general program within a state framework.

Notes

1.   The Committee on Higher Education of the Ministry of Science, Higher Education and Technology Policy of the Russian Federation, directly controls 178 of the 534 higher-educational institutes. The rest report to other ministries and state bodies or are non-governmental establishments.

2.   *Editors' Note:* This example of Estonia must be treated with some caution. As an independent sovereign nation, Estonia is in no way unique in providing educational instruction only in its own language. Moreover, courses in Russian are still available in some institutions of higher learning. To create two full, parallel systems, however, could greatly increase the financial pressures on each. The larger questions of citizenship and language in Estonia with respect to the Russian-speaking population are not just questions of education and cannot be judged on that basis alone.

3.   Sociologists indicate that fewer than 30 percent of students study systematically; only one hour and forty-six minutes were used daily by the average student to prepare for classes and do individual work. Forty-one percent of secondary-school graduates entered higher education in 1987; in 1991, only 28 percent intended to continue with their studies.

4.   In order to minimize the damage of future unemployment, it would be necessary to raise the level of professional skills of almost one million people and retrain about two hundred thousand people with higher education within a year or a year and a half. A total of 3.2 million people would have had to have been re-

trained in 1995, and four million would have to be retrained in the year 2000.

5.   In 1991, a number of institutes of higher education and scientific institutions of the Academy of Science served as a base for over forty study and teaching centers, six research laboratories, and fifty joint groups aimed at solving specific research tasks.

6.   The first technology park in Russia was opened by the Tomsk Polytechnic Institute in 1990. There are eighteen science and technology parks in CIS today, thirteen of them in Russia. The Association of Science and Technology Parks is active in maintaining relations with similar organizations abroad.

Chapter Three
Russia

Science in the Post-Soviet Disunion

Yakov M. Rabkin and Elena Z. Mirskaya

The collapse of the USSR has brought systemic changes to the economy and the political system. In the midst of these changes, the consequences for Soviet science and its culture remain to be determined.

With the demise of the Soviet Union, the Soviet scientific enterprise, previously the largest in the world, has been broken into at least fifteen fragments. But insofar as the national scientific community of the former USSR was more than an imposed administrative structure, cultural and cognitive links fostered throughout Soviet history can be expected to survive the process of political change and fragmentation but not without significant change. In other words, the culture of Soviet science may not be bounded by the new political borders. Long-standing relationships between members of the scientific establishment in the former USSR and the formerly Communist countries of Eastern and Central Europe may yet survive. As national borders have changed, many economic links within the former COMECON have become a liability.[1] This may not be true of contacts in science and technology. However, at this point (1995), the frequency and intensity of these links have been drastically diminished.

Russia dominated the Soviet scientific enterprise. It had most of the scientific workers,[2] most of the scientific publications,[3] and most of the basic research.[4] While the changes affecting Russian science are dramatic, they are likely, for these reasons, to be less significant than those experienced by science in the rest of the former Soviet Union and Soviet bloc countries. It is quite unique for scientific peripheries to be so suddenly divorced from the center to which they have been wed for so long.

The periphery science communities relied heavily on Moscow for intellectual and organizational support. They tended to be characterized by vertical ties to Moscow rather than horizontal ties to one another.[5] Most national academies of science were organized initially as field offices of the Soviet Academy of Sciences[6] and were established by Moscow as tokens of the republics' national prestige. They reinforced the "myth" that the USSR was a society of equal members and, thus, they were deliberately supported by the Communist partocracy, both in Moscow and in the periphery. Most sciences in the Soviet republics had an ornamental, propagandistic character reflecting their political origins.

There was never a significant brain drain from the periphery to the center—that is, from the republics to Russia/Moscow. While there were exceptions, overall, the republics retained their own corps of scientists to themselves.[7] These science establishments thus remained available to the republics following the dissolution of the USSR.

Given these considerations, the issues confronting an analysis of post-Soviet science are different for the other republics than for Russia. The various national sciences may have to maintain links with science in Russia in order to restructure their links with world science, previously mediated by Moscow, and to establish new links with scientific institutions in their geocultural habitat (Turkey, Iran, India for the southern republics, and Central Europe for Ukraine and Belarus). However, many of these new links are taking place without Russian intermediation. Moreover, this issue may become irrelevant insofar as the socioeconomic crises in the newly independent republics have destroyed much scientific activity in the republics.

For Russia, science faces a different set of issues. What mainly concerns Russia is the precipitant decline of the status of science as a key priority for the state. This has had immediate repercussions for funding, recruitment, and, even though political restrictions are no longer in place, links with world science. These changes will have an impact on the unique science culture that developed in the course of nearly three centuries in Russia. This article will focus on the post-Soviet evolution of Russia's science.

Cultures of Soviet Science

Soviet science evidenced several "cultures" of science. At one extreme was a commitment to Mertonian values and a strong motivation to do research for both intrinsic and extrinsic reasons: to leave a mark on world science and to contribute to the welfare of the country. At the other extreme was an interest in science as a source of prestige and financial benefits versus research activity per se. This was reflected in such ways as research done by subordinates being attributed to administrators of research institutes and other members of the ruling Communist class—the *nomenklatura,* who were often unrelated to science altogether. A scientific degree offered both symbolic and material capital as well as a relatively secure place of refuge for members of the *nomenklatura* wishing to hedge their political bets. While the whole range of scientific cultures could be found in the Soviet Union, the essentially artificial character of science in many outlying areas accounted for the dominance of non-Mertonian scientific cultures in many parts of the Soviet scientific enterprise.

Soviet science was hierarchically structured; scientific workers situated at different levels of the pyramid led different lives and had different goals.[8] The top of the hierarchy, essentially the scientific echelon of the *nomenklatura,* behaved similarly to the rest of the Soviet ruling class. Moving down from the top, Mertonian values become more apparent. However, the domination of the governing elite led to a significant polarization within Soviet science, so that it was observed that "scientific researchers became marginalized not only from society but from their own social institution—science."[9] The state, in this sense, was a decisive influence on the social relations of science.

Historically, science in Russia has been linked to Western liberal values. Attitudes to science often served as a reliable litmus test to tell a "Slavophile" from a "Westernizer."[10] Modern heirs of the Slavophiles and the Westernizers have picked up the debate basically where the communists suppressed it seven decades ago. For the Slavophiles, science, like Russian literature, is akin to religion—a source of moral values and societal responsibility. Russia became a propitious ground for the "scientistic" propensity to turn science into a total religion, offering spiritual refuge and protection against the tumult and violence of Russia's political and social life.[11]

The communists saw science as an omnipotent force and the

foundation of a new social order. Science was seen as a moral substitute to organized religions. For the masses, scientists were members of an imposed alien order that gave scientists perquisites and resources well above their perceived usefulness. The state remained a protector of science both in material and ideological terms. Despite the negative implications of this link between science and politics, the support of the state enabled scientists to assume a degree of corporate autonomy unparalleled in other Soviet professions.[12]

At the same time, however, the state never trusted the scientists. Thus, for example, the nuclear program (and many others) was placed directly under the control of the secret police.[13] Such pervasive control over the scientific community was exercised precisely because scientific work presupposes a degree of independence and freedom of opinion. A level of anti-intellectual and antiscientist feeling was maintained in the masses, moreover, through the party-controlled press.[14]

Although efforts were made to instill a certain level of mistrust of scientists and academics on the part of the masses, the former viewed the latter quite differently. As a part of the intelligentsia, Russia's scientists share its specific culture, which includes a strong sense of responsibility for the welfare of "the people." Soviet propaganda cultivated this even more and made scientists feel guilty for their relatively good, state-provided life. When combined with a commitment to Mertonian values, this guilt fostered a culture that emphasized selfless devotion to both science and the people in combination with its virtually total dependence on the state. "The people-loving Russian intelligentsia is alive to the people and exists only thanks to the regime it so much abhors."[15]

Because of its position, science offered a vehicle for civic discourse acceptable to the state. Under the guise of scientific issues, certain political issues could be debated in a seemingly objective and incontestable manner. This role is no longer necessary. The new, free political process has made scientific credentials largely irrelevant to decision-making.

As science loses its place as a source of political legitimacy, being replaced by popular support, more is at stake than its status as an ideological icon. The state no longer needs science to cultivate illusions of a better future; privatization is now used to maintain a belief in a better future. Incomplete assimilation of science by Russia's society becomes a crucial liability when the society goes through another bout of enforced modernization. The combination of anti-Western, antiscience feelings, together with the different cultures of science described above, has caused scientists' self-image to undergo rapid change.

Science, Scientists, and Change

Scientists, as an occupational group, embraced *perestroika* more enthusiastically than others; some of them played a conspicuous and even prominent role in promoting this change. Many well-known scientists, long exposed to Western, mainly American influences, served as conduits of change. Ultimately, it is to the credit of the Soviet scientific community that they acted as they did, given the negative consequences for them that followed in terms of reduced state support, loss of their privileged position, and so forth.

Many of the most prominent names associated with economic reforms while the Communist Party was still in total control of all decisions of state were prominent *uchenye,* a word that literally means "learned." Earlier, it was not Kosygin, the prime minister, whose name was associated with attempted industrial reforms, but Liberman, an *uchenyi* and, what is almost synonymous in Russian popular consciousness, a Jew. The Yeltsin government has more *uchenye* among its ministers than any previous government of the country. *Uchenye* increasingly serve as a convenient scapegoat for Russia's problems, be they economic, ethnic, or environmental.

Russian science has declined significantly in its prestige. The government no longer needs science as an ideological crutch.[16] In the West, modern science has become a catalyst for technological change. This role was largely absent in Russia and, in spite of ideological verbiage, in the Soviet Union. Technological innovation rarely stemmed from domestic scientific efforts; in general, science was a drain on—rather than a contribution to—the country's material resources. Throughout the country's history, science did not become an economic factor as it had in Western industrialized societies,[17] but rather it remained hydroponic, relatively well fed with government infusions but essentially devoid of roots in the mainstream of Russia's culture.

Soviet rulers used to believe that investment in basic research would trigger technological innovation. But by the mid 1970s, the perception changed and research came to be perceived as an expensive activity of little direct economic benefit to industry or agriculture. In reality, however, it was the Soviet economic system that was organically impermeable to technological innovation, particularly from indigenous sources.

Perestroika opened the gates to alternatives that competed with science as the source of truth. Daily horoscopes, the advertisements of healers, and interviews with witches erupted onto the post-Soviet public. While a coherent antiscience discourse is yet to emerge from these disparate phenomena, several scientists have reacted with anger and fear. They see the emergence of these phenomena as a serious danger to science.[18] In fact, it may be a more serious danger to science as a system of beliefs than to science as a research activity.

Survival has become a frequent word in the post-Soviet scientific community. There is talk of "survival strategies" with the caution that "survival is not for all." Government policies suggest that scientists should "save themselves" rather than wait for salvation from the state. Many are leaving science altogether (600,000 between January of 1991 and April of 1992). Many more are maintaining their positions as researchers but get no salary and, certainly, no research funds. Permanent leaves of absence have become commonplace. The gigantic science system of Russia is undergoing a painful slimming program. In the period 1991–94, civilian science funding dropped by a factor of five.[19]

Mechanisms of Adaptation

In the first post-Soviet years, as just noted, the emphasis was on "survival" and the "salvation of national science." Survival options were debated in professional and general periodicals.[20] However, in the years 1994 and 1995 the call for salvation became an anachronism as the awareness set in that the extravagant Soviet science infrastructure could not be "saved" for the future. Several international agencies also concluded that the Soviet system of research was beyond salvation. Within Russia, political support for science and technology is uniform but remains largely rhetorical.

New modes of financing were introduced in post-Soviet Russia.

Peer review and project funding were initially brought in by Soros and other Western agencies.[21] Later, this method was adopted by several state agencies that moved away from exclusively institutional funding typical of the Soviet period. Institutional funding was improved for dozens of State Scientific Centres, the most distinguished institutions that the government chose to preserve from demise starting in mid-1993. However, as the share of salaries in R&D expenditures rapidly increased, spending on new scientific equipment decreased ten times between 1990 and 1993. This trend was particularly damaging since Soviet science had always been underequipped, in both quantitative and qualitative terms.

Gigantic institutes of applied science and technology became deserted as funding dried up, and new activities were often at variance with the declared goals of such institutions. Some were reported to engage in small-scale and low-tech industrial production of consumer items. Others found a source of support in renting floor space to foreign and local firms. Complaints about "unsolicited technologies" proliferated as hundreds of thousands of scientists and technologists attempted to sell their inventions and discovered that there was little, if any, demand for new technologies within the country. The most dynamic economic forces moved away not only from industrial innovation but from industrial production altogether. In 1993 slightly over 1 percent of R&D funding came from "commercial bodies," an imprecise category that lumps together industrial enterprises, banks and trading firms.[22] However, over 12,000 small innovation enterprises (SIE) cropped up in Russia in the years between 1990 and 1993.[23] Some of them benefit from direct state support and all enjoy tax breaks which may in fact explain their remarkable proliferation in the absence of visible innovation activity.

Export of technology initially generated enormous optimism among scientists and engineers. Dozens of projects were launched but relatively few actually succeeded in marketing post-Soviet technologies.[24] Major corporations, such as Boeing and Daimler Aerospace, established a permanent presence in Russia and appear to have benefited from technology transfer.

Foreign participation has become more important in funding basic research. Various estimates suggest that between one third and one half of research funding actually comes from foreign sources.

Precise data may be impossible to obtain since much of the funding has bypassed usual banking channels and has been done in cash. This was a reaction to the instability of Russia's banking system, compounded by high levies on transfers of foreign funds into Russia. International financing of Russia's research brought it closer to world trends in science but may have also depressed the originality of research approaches that used to be protected by the iron curtain. Pendulum migration, that is, time sharing on the part of Russian scientists between a foreign laboratory and his home institution, constitutes another facet of foreign support of Russia's science.

A few new institutions have appeared since 1992. They are leaner, involve theoretical rather than experimental research, and derive financing from Western sources. The Euler Institute for Mathematics in St. Petersburg, for example, hosts first-class seminars but "when the visiting scholars depart, its halls are almost empty."[25] An independent university of theoretical science in Moscow, supported by a few countries, has attracted remarkable teachers and students. This is noteworthy in the context of precipitously falling numbers of applications to the relatively strong faculties of mathematics and physics at Moscow University.

New forms of organization of post-Soviet science have included "science exchanges," one of a series of attempts to market scientific ideas. Banks of ideas have sprung up, but their viability is gravely in doubt. Some have been reported as fronts for transferring state funds into the pockets of private individuals, a common feature of post-Soviet development. A novel idea of saving Russia's science suggests that a portion of Russia's external debt be swapped for funds Russia would spend on research. The results of such research would be made available to both Russia and the creditor nations. However, such ideas have not been taken seriously by Western bankers and policy makers.

New government structures have been established to regulate the largely chaotic transformations in Russia's post-Soviet science. The Ministry of Science and Technology Policy has assumed an overall responsibility for R&D in the country that brought it into open conflict with the Russian Academy of Science, traditionally "the headquarters of the nation's science." The Foundation for Basic Research has become the vehicle for project financing and has reached ca. 2 percent of the state R&D budget.[26] Another administrative innovation of President Yeltsin was the Fund for Technological Development started in 1992 to channel a part of state taxes on industrial sales to applied research and technology.

Nongovernmental structures and associations have also been established, and many have become active as lobbying groups. One such group, the Association of Science Sites, defends the interests of dozens of remote research centers that have been dispersed throughout the country by successive Soviet governments and are currently facing acute crises compounded by their geographical isolation.[27] Professional societies, such as the Physical Society, also have proliferated, some of them encompassing several former Soviet republics.

The contraction of the research system and the sudden loss of prestige and privilege emphasize the fragility of science in Russia. "In our country, science can exist either as an *affaire d'état* or it cannot exist altogether."[28] This opinion, which echoes the more conservative analysts in the immediate post-Soviet years, is nowadays shared by a broad spectrum of professional and political opinion, both in Russia and abroad. Structural and cultural changes have taken place in Russia's research system, bringing the system closer to Western models. While the flux in the country's economic and political development makes it impossible to fully assess the future of Russia's science, it appears that the adaptation mechanisms, for now, are keeping the crisis in check.

Acknowledgment

The authors acknowledge advice on an earlier version of this chapter received from Spas Spasov and Adel Ziadat.

Notes

1. Françoise Nicolas and Hans Stark, *L'Allemagne: une nouvelle hégémonie* (Paris: IFRI, 1992), pp. 119–20.
2. By 1990, more than two-thirds of the 1.8 million Soviet scientific workers were located in Russia, with Moscow the center for the entire union. L.E. Mindeli, ed., *Nauka SSSR v Tsifrakh* (Moskva: Mysl, 1991), p. 17.
3. In 1989, for example, Moscow produced 40 percent of the country's internationally visible publications, five times the level of the next closest center, Leningrad (St. Petersburg).
4. By some estimates, as much as three-quarters of Soviet fundamental research was located in Russia. I.V. Marshakova, *Sistema tsitirovaniya* (Moskva: Nauka, 1988).
5. To an extent, the networks in Soviet science could be compared with those

linking anglophone Africa with Britain in the 1960s and 1970s. Thomas O. Eisemon, *The Science Profession in the Third World: Studies from India and Kenya* (New York: Praeger, 1982); Yakov M. Rabkin, et al., "Citation Visibility of Africa's Science," *Social Studies of Science*, 9 (4), 1979, pp. 499–506.

6.   Yakov M. Rabkin, "Academies: Soviet Union," in *Encyclopedia of Higher Education* (New York: Pergamon Press, 1992), pp. 1049–55.

7.   Moscow would only seldom offer comparable living conditions for members of the republics' scientific elites, who enjoyed ample, sometimes extravagant, perquisites and privileges on their home territory.

8.   M.G. Yaroshevsky, "Stalinizm i sud'by sovietskoi nauki," in *Repressirovannaya nauka* (Leningrad: Nauka, 1991), pp. 9–32.

9.   Elena Z. Mirskaya, "The problem of Justice in Soviet Science," paper for the workshop "Science and Technology with a Human Face," Moscow, September 1992, p. 13.

10.   Alexander Vucinich, *Science in Russian Culture* (Stanford, CA: Stanford University Press, 1970), vol. 2.

11.   David Zhuravsky, "Neosuschestvimyi prockt Ivana Pavlova," *Voprosy frosofii*, 9, 1991, pp. 27–29.

12.   Yakov M. Rabkin, "Scientific and Political Freedoms," *Technology in Society*, 13, 1991, pp. 53–68.

13.   Andrei Sakharov, *Memoirs* (New York: Knopf, 1990).

14.   It was commonplace to find reports of misdeeds on the part of individual scientists alongside prose praising Soviet science in general.

15.   E.V. Semionov, "Nauka protiv proizvola," paper for the workshop "Science and Technology with a Human Face," Moscow, September 1992, p. 14.

16.   Reagan's Strategic Defense Initiative (SDI) in 1980 shook the confidence of Soviet leaders in national science; it may even be argued that the SDI triggered *perestroika*. Gorbachev realized that military competition with the West was no longer feasible or affordable and the *nomenklatura* began to assimilate Western approaches and attitudes with the express purpose of perfecting rather than abolishing the old system.

17.   Yakov M. Rabkin, "Transnational Invariables in Science Policy," *Administration publique au Canada*, May 1981, pp. 18–43.

18.   A good example of this reaction was offered by Sergei Kapitsa, a well-known physicist and popularizer of science, who vehemently denounced these novel alternatives to science in the course of a seminar on antiscience movements in the United States and the USSR held at MIT in May of 1991.

19.   O.V. V'uugin et al., *Rossiya: ekonomika i nauka na puti reform* (Moskva: TsISN, 1993), p. 69.

20.   Igor Troitski, "Bei Doktorov—spasai nauku?" *Radikal*, 41, 1992, p. 3.

21.   International Science Foundation, *The 1993 Annual Report*, Washington, DC: ISF, 1994.

22.   N. Gaponenko, "Transformation of the Research System in a Transitional Society: The Case of Russia," *Social Studies of Science*, 25 (4), 1995.

23.   *Nauka Rossii v tsifrakh: 1993*, Moskva: TsSn, 1994, p. 23.

24.   Yakov M. Rabkin, "Organizatsia sovmestnogo tsentra po peredache tekhnologii," in *Nauka i tekhnologiya: Rossiya i mir*, Moskva: Minnauka, 1994, pp. 144–50.

25.   "Storm Clouds over Russian Science," Science (Special Report) 264 (May 27, 1994), p. 1276.

26.   *Vestnik RFFI* (Moskva: RFFI, 1994).

27.   Natalia Nikitina, paper for the NATO Advanced Research Workshop, Nice, March 1995.

28.   E.V. Semionov, *op. cit.*, p. 23.

Chapter Four
Science, Technology, and Higher Education in the Baltics

A.D. Tillett and Barry Lesser

The achievement of independence from the Soviet Union by the Baltic states of Estonia, Latvia, and Lithuania in 1991 marked the culmination of some fifty years of occupation. But independence brought with it a host of problems. Some of these problems were a function of the process of political and economic liberalization that accompanied, and in some areas even preceded, independence. Others were a function of the need to reorient and adjust Soviet institutions and systems to better meet national goals. The educational and scientific establishments in the three Baltic nations faced problems from both of these sources. The challenge they now face is how to maintain their science and technology infrastructure while bringing it into line with the new economic and political realities in a manner consistent with national priorities and their newly regained status as independent nation states.

As part of the larger integrated system of the USSR, the science and educational systems in the Baltics were designed to serve the needs and priorities of the larger system. In organizational and structural terms, they mirrored the system they belonged to, both in their strengths and their weaknesses. When they were divorced from the larger system, they faced the task of long-term structural reform, along with the short-run task of coping with the new economic environment, which was one of restraint, reform, and restructuring. Finding a way to meet the long-term task and convert their science sectors into national assets requires a long view, which may not be consistent with the economic and financial realities they now face. Finding a way to reconcile these two views will not be an easy task.

The Changing Role of Science Institutions
The three Baltic nations see their national science establishments as

assets. But, equally, they recognize that changes are necessary. What is not certain is what, ultimately, these establishments will become.

There are a very few individuals who would prefer things to remain as they were before independence. There are more who recognize that change is inevitable, and perhaps necessary, but who want to proceed slowly, both to protect their own positions (such as researchers afraid of losing their jobs or status) and because they believe that the issues are too important to risk acting in too quick and precipitate a manner. There are those, also, indeterminate in number, who believe that these small, relatively poor countries cannot afford to maintain significant science establishments of their own; the myriad of other social and economic problems they face as part of their adjustment process simply reinforces this view. Finally, there are many who believe the science system must receive the resources necessary to maintain itself in the short run, while the changes, or debate about the changes, proceeds. Otherwise they fear that the system will be irreparably damaged before its future course can be determined and implemented.

There are three principal types or sources of pressure on the system:

1.   Resource pressure—the shortage of funds and the excess of needs into the foreseeable future.
2.   Economic pressure—the requirement for science to assume a role that makes an explicit as well as intellectual contribution to social welfare.
3.   Civic pressure—the requirement that institutions (and their personnel) be more responsive, transparent, and forthcoming regarding their activities, values, and directions.

For these pressures to be met, science and its associated enterprises will have to enter the public arena more fully and demonstrate their potential contribution to building the new society.

In the Baltics, as elsewhere in Eastern and Central Europe, discussion encompasses both values and mechanisms. For example, there is relatively general agreement that research should be more fully integrated with teaching—that is, more fully integrated into universities.

The division of teaching and research institutions characteristic of the Soviet system is seen to weaken both, when rigidly applied. Although a number of research institutions may well remain independent, their staff should involve themselves in university activities at least on an honorary or advisory basis. There is less consensus regarding a different but insistent point of discussion, that of science as an income earner. Many feel that the work undertaken in scientific institutes should command a satisfactory premium from local industry or enterprises; many others, not only scientists, hold the view that knowledge has value in itself and that the commitment to universal knowledge should be demonstrated by governmental support of science as an intellectual enterprise. This view may seem somewhat quixotic but few could or should despise it, because there are few countries in which learning has been so important a value in preserving a historic sense of nationality.

The important question then is: Given that these states have large scientific systems that must change, how will the values of education, income, and knowledge be translated into priorities and choices? Perhaps the most radical view is that market principles should apply to research and that governmental subsidies should be withdrawn from all but a minimum of basic science institutions. The remainder would earn their keep by selling their wares. However, it is difficult to consider that the industrial or agricultural sectors are yet mature enough to develop a private enterprise structure able to finance a great deal of applied science or technology. Certainly, to assume that a science market will come about easily is to overestimate the nature of the current science on sale. Further, market-driven science is an exception and is typically the result, as in the case of biotechnology, of large, developed infrastructures. The inverse of this criterion is that of science quality, a judgment that a science project or activity has the capacity to make a contribution to world knowledge and should be preserved. There is clearly good and bad science, and it is most useful to know, as will be discussed below, how a particular branch is ranked by experts and practitioners. Certainly some proportion of resources should be allocated on this basis not least because, in a number of fields, there are few apart from scientists who can make an informed judgment. It is unlikely, however, that this could become the only allocation principle and therefore, like market principles, it will have to be modified.[1]

A third approach, which does not by itself solve the overall problem of the future role of science, is to look for foreign support in the form of cooperation, contracts, or grants. However, given the size of the research establishment and diminishing resources, few countries can avoid the unenviable task of *picking winners*—that is, deciding on priorities by taking into account as much information as possible. For this approach to receive support, allocation must take public preferences into account. Today, elected officials set public preferences, often line item by line item, so that scientists, sooner or later, become lobbyists for their cause.

The changing role of science is a complex process and, although perhaps marginal to the short-term economic stability of the Baltics, it is recognized as a key to their future economic growth. What makes up a science system and how is a proper balance between basic and applied, pure and commercial, research and teaching, to be arrived at and maintained? How are these countries dealing with these seemingly intractable issues?

Lithuania

As of 1992, higher education in Lithuania consisted of six universities and ten academies. Scientific research was divided among twenty-nine state research institutions and affiliates and fifteen state scientific institutions. The distinction between state institutions is apparently historic, with state research institutes forming part of the previous academy structure. As of 1993/94, these numbers had changed to fourteen institutions of higher education (universities and academies), twenty-nine state research institutes, and twenty state scientific institutes.

The Law on Science and Studies, which came into force in early 1991, established the principle of autonomy for both institutes of higher education and state research institutes. Universities have elected senates, which in turn elect the rector. State research institutes are also autonomous once their statutes, like the institutes of higher education, have been approved by the Supreme Council (Parliament). State scientific centers are not autonomous and their heads are appointed by the relevant ministry or department.

The policy structure consists of the Lithuanian Agency for

Higher Education, Research, and Development, the government's policy and administrative agency; the Science Council of Lithuania, which establishes priorities and supplies and allocates the science research budget; the Lithuanian Academy of Sciences, which is now an academy of distinguished scientists honored for their contribution to science; and the Lithuanian Union of Scientists, which acts as a combination professional association and information agency. There is also a Rector's Conference of Lithuanian Higher Education Institutions.

The decision to relieve the academy of its administrative role was one of the first breaks with the past; the key institution now would appear to be the Science Council, which is responsible for distribution of research funds provided through the Lithuanian Research and Higher Education Fund. There is also an Innovation Fund to finance applied research in new technologies. The council consists of thirty-six members, two-thirds of whom are elected by the scientific community and one-third appointed by Parliament. It has three commissions: a finance and priorities committee that searches for ways to raise funds; a committee that is examining the qualifications, mandate, and organization of the different state institutions; and a committee on informatics. The full council meets every two weeks, the different commissions once a week.

Financing is widely regarded as the key issue currently facing the scientific community. Budgetary allocations are obtained directly from Parliament so that the position of chief scientific adviser, who is also pro vice rector of Vilnius University, becomes very important. This person chairs a board of ten members consisting of representatives of the Lithuanian Academy of Sciences, the Science Council, and the Rectors' Conference. He is partially responsible for persuading the government to provide funds for science and undertaking negotiations with the relevant parliamentary committees. The scientific budget as an independent line item is debated by Parliament. In 1994, 6.1 percent of the state budget was allocated to research and higher education.

One of the most troublesome issues facing the scientific community is salary levels. The council is hoping to have salaries indexed because of their concern that the brightest and the best are moving away or going into private business. There is also concern about the age of research scholars and the importance of supporting junior scholars. There is a strong sense that many respondents regard the decline of the scientific community as regrettable but inevitable.

Efforts are being made to develop a set of priorities and an overall operational framework that will place a priority on issues that concern Lithuania and can be studied only by Lithuanians—such as history, language, and literature—as well as environment, agriculture, and geology. Part of this exercise includes an examination of the sciences considered to have the best possibilities of success, such as mathematics, statistics, physics, and chemistry. Some respondents felt that Lithuanian science activities should concentrate on a combination of newer fields of learning for Lithuania, such as microbiology, economics, and languages, and avoid areas of specialization where other, particularly Eastern European and former Soviet countries, have an edge. In addition, the council is looking at personnel qualifications and the development of training schemes for requalification.

A second key activity is a review of state research institutes funded directly by the government. The institutes were given a two-year lease on life in 1992, after which time a review was to occur regarding their continuation, liquidation, merger, or integration with the universities. A number of institutions which were strongly associated with the military have been or are to be sold. It is recognized that the review will have to be undertaken carefully and could lead to extremely tense personal relations. It is confidently assumed that institutes of law, sociology, and philosophy will be integrated into universities. It is clear that not all the state research institutes will survive, at least in their present form. In 1994, the review process was initiated with the formation of a Coordination Commission. Institutions undertook an initial self-assessment and will subsequently be evaluated by two peer groups formed by the Science Council and the Academy of Sciences.

A third important task is that of supporting the integration of research institutes and universities, particularly as there is no ministry of higher education although the latter was proposed in 1994. The integration of science, it is recognized, must also be accompanied by a sense of the value of particular specializations—the new university of

Klaipeda, for example, will emphasize machine building, while Kaunas Technical Institute is expanding as a means of linking various medical and technical institutes. Scientific work is now expected to help resolve national issues such as the environment, particularly in relation to energy. And scientific knowledge is important to the resolution of various practical problems, such as the Ignalina nuclear plant and expansion of the oil terminal at Klaipeda.

The change in university governance has been accompanied by an attempt to reduce class size and formal meeting times. The integration program will permit universities to offer upgraded programs, and several commentators have mentioned the importance of expanding these programs to accommodate more students, possibly double the present number. Officials believe that links with foreign universities and institutions will help revitalize Lithuania's universities and so promote contacts and agreements. The European Community is lending technical assistance to universities through the TEMPUS program, and there is greater contact with the non-Russian world by means of the publication of magazines in English.[2] Significantly, the number of students in institutions of higher education has been falling steadily since 1989.

Overall, Lithuania's scientific choices have yet to be made. It is widely believed that the system will need at least ten years before it becomes stable and fully productive. In the meanwhile, the process of change must proceed; the status quo is no longer defensible.

Latvia

Latvia's institutions of higher education and research are also attempting to overcome the legacy of the USSR. There are now seventeen institutions of higher education (universities and academies). Such institutions undertook little if any research under the Soviet system, while the institutes of the Academy of Sciences, some of which were very large, concentrated on pure science in cooperation with thirteen state enterprises. The USSR was the key to science and the Baltics were in general left with fundamental sciences; there was little connection or encouragement to link with state companies, so that science was kept as a largely independent enterprise. Scientists who came from the USSR, according to one respondent, were of poor quality, and their presence (they make up about one-third of the professors who have stayed at the universities) is complicated by the citizenship question, an issue that dominates Latvian politics and policy.[3]

The Latvian authorities have introduced a number of changes that will affect resource allocation. First, universities have become autonomous and are being encouraged to review and update their curricula while expanding their research capacities. Second, there has been a reconfiguration of administrative responsibilities between the different scientific bodies. Third, to confirm and promote these changes, Latvia has introduced a competitive grant system. And finally, with the vital assistance of the Danish Research Councils, there has been a major assessment of all scientific activities in Latvia.

Universities   Universities have now established their own boards and are writing new regulations, reviewing curricula and entrance requirements, and validating the qualifications of teachers and graduates. There are now five universities in total. The two most important universities in Latvia, the University of Latvia and Riga Technical University, are line items in the national budget and have direct contact with government and Parliament. There are key changes underway or under consideration that are likely to alter university performance:

1.   The revision of the degree system in order to accommodate baccalaureate, master's, and doctoral degrees, a hybrid U.S./U.K. system. Doctoral candidates will no longer have to pass examinations in philosophy (principally Marxism), a language (normally Russian), and a chosen subject area. Rather, the examinations will be broader and directed toward other languages while maintaining a central role for Latvian. The specifics of these decisions will be determined by the autonomous universities themselves.

2.   Establishment of research centers within universities as the previously independent research institutes are encouraged to associate and become part of an institution of higher education. There have been some examples of this in the humanities and social sciences, but the most significant to date is the move of the Institute of Molecular Biology to the University of Latvia.

3. The validation of faculty degrees, previously awarded by the USSR, and now to be validated by Latvian authorities. The validation process consists of (a) the provision of the previous dissertation, (b) a description of the results in the Latvian language, and (c) a list of papers. The Soviet degrees are not so much being challenged as an indigenous, Latvian system is being promoted. But a professor in the Latvian system must complete this process. Each university has set up its own boards—there are eighty such boards at the Technical University, for example—to carry out this validation task, which will be available only during the transition period.

Scientific Reorganization The Latvian Council of Science, established in 1990, is a key institution for setting science and education policy. It prepares the research budget for submission to the Council of Ministers, allocates resources, and coordinates activities among the government, research institutes, and universities. The council consists of twenty-eight members, twenty-four of whom are elected by scientists. The council organizes fourteen expert commissions that judge project applications.[4] The expert commissions are elected by scientists, in 1991 involving 174 scientists from different parts of the system. The Latvian Academy of Sciences (LAS), which used to be responsible for administering seventeen research institutes through three divisions, is now converting itself into an institution of leading scientific experts. The role of the Academy is no longer to administer institutes, which are described as "practically autonomous government financed institutions,"[5] but to promote scientific activities.

Science Funding A key alteration, which reinforces the organizational and educational changes, is the move from institutional to grant competition. The government budget, which is negotiated through the Latvian Council of Sciences, is divided between infrastructure (30 percent) and project grants (70 percent). Grants are awarded on a competitive basis and judged by the council's expert commissions; in 1991, when the scheme commenced, 830 research projects received funding from a total of approximately 1,000 applications. The grant is held personally and the total amount is prescribed in the following way: salary 52 percent; social security, about

19 percent;[6] infrastructure, 20 percent; and 9 percent for research materials, travel, and other related activities. Grants are small and inflation has eroded them further, but the council intends to make grants the principal method of science funding in Latvia. As of 1993, the Department of Higher Education and Science was established within the Ministry of Education, Culture, and Science; the department oversees the science funding system, among its other duties.

Grant funding has important implications for the research institutes, universities, and Latvia's science future. First, institutes will no longer be supported by block grants but will have to compete for the 20 percent of the individual scientists' grants earmarked for infrastructure. Institutes will remain independent if they can attract a significant accumulated portion of infrastructure awards; scientists will then sign what appear to be service contracts with their host institution. In general, council officials wish to encourage research institutes to associate with universities but are willing to let a science market develop and to close those institutes that are unable to attract scientists with grant funding. The running costs of some of the leading institutes have proven expensive. In 1992, for example, the Institute of Physics consisted of six or seven buildings but fierce winter heating costs forced members to move into one building, demonstrating that, in order to survive, institutes may have to become both more efficient and smaller. A number of the laboratories have since moved to the university, and total staffing has declined to two hundred from a previously estimated eight hundred employees.

These changes are clearly difficult for some of the institutes. If they do not receive a grant or are unable to move to a university or other institute of higher education, staff have gone to half-time salaries or become underemployed. Such a circumstance is particularly serious for older scientists; those under thirty might expect to receive grants from abroad or have greater alternative employment opportunities.

The new approach has, of course, aroused some controversy. The Latvian Academy of Sciences, for example, has commented that the new scheme

did not work the way it was intended. The most capable young scientists leave research institutes and go into industry. It seems that the present crisis in the national economy and other spheres of life undermines the prestige and confidence in the future of Latvian science.

There must be some question whether the market approach to institutions can be completely carried through. On the one hand, the scheme has shaken up a rigid and static system, not least in reducing the number of applied research institutions, some of which were associated with military production. The government, through the Ministry of Education, is providing some funds based on joint applications from industry and science, provided that the application meets a number of commercial and technical criteria. On the other hand, salaries are low and the government has repeatedly reduced overall funding. Under present circumstances, although commitments are made for three years, money is guaranteed only for one year.

As the report of the Danish Research Council shows, there are many aspects of Latvian science that are worth preserving. The report, published under the title *Latvian Research: An International Evaluation,*[7] is a review and evaluation of all scientific projects being undertaken in Latvia. The work, which was completed in 1992, was based on nineteen different panels involving about eight hundred different projects; the evaluations were based on written work and site visits. The panels examined projects in terms of scientific quality, group or institutional capacity, and the potential value of the work for Latvia. The panels were also encouraged to comment on scientific and administrative staff, accommodations and equipment, financial conditions, and their relationship to higher education, and also to make recommendations regarding the future.

The general conclusions of the report can be summarized as follows:

1. Latvia should maintain "a balance between the long range basic science and short range applied science . . . remembering that the foundation of a part of the future development of Latvian society is created by the activities of a stable high quality research base"; special support is recommended for fundamental medical research.
2. Measures should be taken "allowing the survival of top class research identified by the evaluation programme in order to protect

and preserve talent, skills and experience, the best guarantee for the development of new directions of research at a high, internationally competitive level."

3.   Research and higher education should be integrated, "but emphasize that this should be carried out gradually in order to avoid unnecessary tensions and to protect well functioning units. The integration should be made in a way ensuring the recruitment of young scientists."

4.   Stronger links should be created between agricultural research institutes and state and private farms.

5.   The isolation of Latvian science should be broken by the development of foreign-language skills, particularly English, publishing in international journals, linking to international computer networks (such as Internet), and providing foreign currency for journals and study visits.

6.   Technology transfer should be encouraged.

The evaluation teams were concerned about the current financial situation of scientists. They stated that "unless immediate measures are taken to ensure a stable funding of the research system, including salaries at a competitive level and funding of the infrastructure, the future existence of research in Latvia is endangered." They appeal to the international scientific community for support, as well as proposing cooperative agreements for contract research with international and other institutes.[8]

     The report, which confirms the importance of many scientific areas, may have made the government's task more, rather than less, difficult, in that it implies that certain areas should become protected species. In the long run, however, the evaluation makes a major contribution to science planning, because it provides a public basis, in many extremely difficult and expert fields, for discussion and choice. It demonstrates the need, difficult in itself, to provide guidelines for scientific and educational research, around which there is a consensus that what is desired is a combination of quality science and applications within the context of national needs. Science officials believe that there are a number of key areas, such as biotechnology, new materials, and informatics, which Latvia must develop. They also believe there are a number of national priorities, such as energy, food production, and alcohol abuse, that require immediate attention and that are now major constraints on economic and social development. Finally, they identify a set of longer-term national issues, such as Latvian language and history as well as the need for research to counter past Soviet damage to the environment and to preserve and understand the nation's ecology.

The scientists who run the principal institutions understand the need for a science strategy tailored for Latvia's circumstances—"to find a niche and be original," as one commentator put it. They recognize that a growing proportion of science activities must be linked to both universities and industry and that scientists will have to become active in national discussions over such issues as nuclear energy, and demonstrate their value to politicians and the public. There is a strong commitment, at the top, to a mechanism that could lead to quite radical change and that is not simply a mechanistic repetition of other models as opposed to an organic response to Latvia's needs and resources. Gaining public and political acceptance of these views and successfully translating them into action by overcoming the forces of intransigence, inertia, and opposition remain formidable challenges.

Estonia

Estonia is moving cautiously toward a science and research system that integrates university research with the work of its academy's research institutes. It is also putting in place nine extrabudgetary funds, accountable to different governmental ministries, intended to promote scientific research, innovation, and the wider use of informatics.[9]

There are six universities, now autonomous, and a number of other colleges or academies that award specialist postsecondary diplomas. The Estonian Academy of Sciences has continued, until recently, to be responsible for research institutes which, until last year, were regarded as different scientific divisions within the academy.[10] They are now divided as research institutes (19), self-supporting institutions (6), and the central administration (3).[11] The government has established a Science Council which recommends and administers grants approved from the Science Fund, which is part of the budget of the Ministry of Culture and Education. A draft law introduced in 1994 would transfer all academy institutes to the jurisdiction of the Ministry of Education; the academy would become a personal institution of distinguished scientists (scholars). Also, as of 1994, discussions were underway on creating a University of Tallinn, a move which, if successful, would further reduce the number of independent universities.

Two major forces determining Estonian science policy are (a) a commitment to educational change, in which science and research have come to be regarded as part of a larger, essentially social problem to which a renewed scientific profession can make an important contribution; and (b) the government's wish to make science more relevant to Estonian industry and society. These social and economic imperatives are overshadowed by major economic constraints that have a continuing and growing impact on the way science is to be funded and organized. A small commission is currently writing a new law on science, described as being "urgently needed," and which is expected to create a General Council of Science and Higher Education, bringing together the various parties involved in science and technology policy. There is some concern within the science community that the baby could be thrown out with the bath water—that the new law and structure might end the gradualism that has been characteristic of the structural changes within Estonian sciences to date. A new science policy will attempt to break down the excessive specialization or "compartmentalism" that was characteristic of the Soviet system and that hinders a more efficient and creative use of resources in Estonia.[12]

Universities A 1993 Finnish report diagnosed the principal constraints on higher education as a lack of policy and budgets, weak university administration, irrelevant teaching and research, and a failure to come to grips with the current needs of the Estonian economy and society.[13] The principal universities involved in science reconstruction, Tartu and Tallinn Technical, are not faced with science issues alone, although they form an important part of their new mandates. Rather, they have to renew and revitalize their present structures and to that end are undertaking three different tasks, any one of which would be a challenge in itself.

First, the universities are commencing research activities—not a traditional role in Estonian universities under the Soviet system—by attracting members or sections of different research institutes to universities. For example, in 1992, two research groups moved to the Estonian Agricultural University, the Institute of Biophysics moved to Tartu, and the Institute of Astrophysics to Tallinn Technical University. Further, a number of joint chairs and cooperative activities have been initiated to link the universities to the academy. Second, the universities are reviewing their curricula and introducing a number of subjects—such as public administration—into their work. Tallinn Technical University (TTU), for example, wishes to expand its range of subjects, partly to move from the Soviet model of a technical institute and partly in response to student needs. Between 1990 and 1993, ten new departments were created at TTU and over fifteen at Tartu. High-quality teaching materials need to be developed, preferably in Estonian, and teachers will have to be retrained. Third, universities are also concerned about the quality of secondary education with its emphasis on technical and specialized subjects.

The present rector of TTU describes the long-term goals of his university as:

linking the education practice to training and research; cross disciplinary research with a systems view, active involvement of industry personnel, university staff and students, to improve industry/university relations; course and curricula development to meet the needs of labour market and to give students certain freedom to influence the content of their studies; a strong commitment to technology transfer.[14]

There is a danger that, far from stimulating students, the combination of changes and uncertainty will prolong the students' learning passivity—a legacy of Soviet-style education—and that, what with increasing economic difficulties, such as the reduction from full to partial payment of student stipends, students will be discouraged from attending universities. These changes could, as university authorities recognize, sap any enthusiasm for reform.

Governmental Policy The government continues to be the principal source of funds for science. Government officials admit that they have yet to develop a firm direction, although the outlines of a new policy are becoming apparent and are likely to be contained in the new law. The initial steps have already been taken to alter the dominant role of the Estonian Academy of Sciences (EAS), encourage research centers under their aegis to transfer to ministries or the universities, and alter the structure of funding from direct support to competitive grants, with the creation of the Estonian Science Fund (ESF). The ESF is intended to support basic and strategic research in seven areas; it is financed from the government's general science and education budget, voted by Parliament, money that previously went directly to the EAS. In 1992, the ESF distributed 5 percent of the science budget, and in 1993 the proportion increased to 25 percent. In consequence, the budget for the EAS and its associated institutes was reduced by 5 and 25 percent respectively. The change not only has a major financial impact on the EAS but also alters the unit of accountability from research institutes to individuals or groups of researchers. In addition, the government has created an informatics fund to encourage the broader use of computers and information systems, and an innovation fund, managed through the Ministry of Economic Development, to promote links with industry and the private sector. These funds are intended to provide incentives for change, although there must be some question whether, given current economic conditions, the amounts are large enough to achieve this end.

Academy and Institutes The Soviet science structure is slowly being dismantled and with it the role and function of the Estonian Academy of Sciences. The academy's 1992 annual report identifies four major changes: first, a concentration on key projects and the rearrangement of research units; second, greater autonomy for the institutes;[15] third, increased integration between institutes and universities; and fourth, encouragement of increased foreign contacts.

The academy has seen its budget and number of personnel steadily decline in recent years. The draft law of 1994, if passed, will complete the process of converting it into a private academy.

Science Evaluation Neither current funding nor the number of researchers is a helpful guide to the future. Accordingly, the Estonian authorities arranged for an evaluation of the natural and social sciences to be carried out by the Swedish research councils in 1992.[16]

The *Evaluation of Estonian Research in Natural Sciences* was undertaken in fourteen research areas, reviewing written material (published and unpublished) followed by site visits. The result was an extremely valuable document that provided crucial guidance regarding the quality of science in Estonia. The projects were judged by international standards, although comments on the individual sectors demonstrate a considerable and sympathetic understanding of the obstacles faced by Estonian scientists today. The Swedish scientists write:

There are no doubts that many talented people have never had the chance to show what they could have done given proper opportunities—and that even the really good projects and the most internationalized researchers (and there are indeed a good number in both categories) have suffered from partial suffocation. Estonian science will now have to function within the competitive and increasingly integrated modern scientific world, with its many international programs, where influence is mostly related to the relevance and quality of one's own input.[17]

The report advocates a much stronger integration of research and education; the importance of encouraging younger scholars in particular because of the greying age profile of Estonian scientists; a simpler and less cumbersome administrative system; and the vital importance of internationalizing Estonian science. In this regard, they advocate the importance of the English language for training and publications; improving scientific libraries; changing the pattern of scientific collaboration from Russia to other countries; supporting foreign exchange programs; updating the communications system; and "Nordification," a deliberate policy of linking with neighboring countries that have somewhat similar scientific traditions.

The specific recommendations—by area, institution, and project—can only help the government and the Estonian scientific community plan a more rational science policy. However, it is possible and recognized that without an improvement in economic conditions, Estonian science will be reduced in size even further and some of the best scientific groups dispersed and lost altogether.

Conclusions

The most important feature of the changes in the sciences within the Baltic states is also one of the most encouraging. Science has become public property, an issue of open debate and partisan discussion, of national choices with important consequences. If scientists feel less important—and thus perhaps regret this change—it is because the hidden world of the *nomenklatura* has been replaced by the scientist as citizen. It follows from this change that the most direct impact of the changing structure of science will be in education rather than research. A more open and enlightened science, tempered by the recalcitrant world of the classroom, can only benefit civic goals.

This feature should not be overlooked even though it may not be immediately beneficial to those facing the current crisis. There is no easily foreseeable way in which to reach this civic goal; science institutions, like all organizations, will have to go through a wrenching process of change. Science can no longer claim a special status and this is all to the good if, at the same time, it becomes more responsive to the society that pays its bills and encourages its activities. The issue of public trust will become an important test for science that an association with education will enable it to pass more easily.

The first steps in dismantling planned science have now been taken by the three governments and their scientific communities; they have involved the development of cooperative protocols between teaching universities and research universities that can only benefit both; a growing realization that the science infrastructure is too large and inefficient to maintain; and a willingness, in different degrees, to place the responsibility for national scientific progress with scientists themselves. The approaches differ by country and by institutional configuration, and it is likely that, in ten years, each of the countries will appear even more distinct. Much will depend on each of the Baltic states' ability to establish a viable and stable economy.

One recent development of note is a movement toward cooperation among the three Baltic countries in the area of quality assurance in higher education. At a meeting in Riga in October of 1994, the three countries signed a Declaration of Intentions on cooperation for quality assurance in higher education in the three countries. In the declaration, each country pledged to create a national institution for accreditation and an independent national body for quality assessment of higher education; to initiate a pilot program in this regard in 1995; to cooperate with each other in this effort by establishing an independent body to coordinate standards, procedures, and mechanisms among the national bodies, to cooperate in training experts for peer review, to share knowledge on individual experience, and to work toward harmonization of standards, procedures, and mechanisms; to cooperate in regard to admission standards for institutions of higher education; and to cooperate over recognition of higher-education credentials issued by the institutions in each country. This initiative stemmed from a regional advisory mission of the Legislative Reform Program of the Council of Europe, in response to a request from the three Baltic states.[18]

The Baltic states face difficult choices regarding the economics of research. For both intellectual and economic reasons, their scientists must view their economic futures as linked to those of the Nordic, European, and Western countries, as their policy-makers already do. Improved equipment can compensate for the loss of personnel, and electronic systems, computer data bases, and conferencing will provide a vital and immediate link to the world of science and learning. Further, the important science communities of the Nordic countries, themselves relatively small but with significant scientific achievements, demonstrate as well as those of any other countries both capacity and creativity in their relations with the world. Recent history shows that it is better to be small and intelligent than the opposite.

The three Baltic countries can rely on important expatriate communities that can be expected to assist their national science communities as they have helped in other sectors. The European community has already begun to open up some of its programs, like Tempus and Eureka, and it is to be hoped that they will continue to support the rapid and pressing changes required of Baltic higher education. North American institutions should also be encouraged to link with Baltic science institutions and universities in their technical assistance programs, in order to encourage both short-term training and the pursuit of postgraduate degrees. Finally, it is to be hoped that the European Bank for Reconstruction and Development will not avoid this area, concentrating on private sector initiatives and balance of payments support alone, but will attempt to find suitable projects to ensure that Baltic science and education will continue to be linked to international science and learning.

Notes

*An earlier version of this paper appeared as A.D. Tillett, "The Baltic States: Higher Education and Science," in Klaus Hüfner, ed., *Higher Education Reform Processes in Central and Eastern Europe*, (Berlin: Peter Lang, 1995).

1. For an interesting discussion with relevance to science policy in these and other states, which shows the value of pluralism and multiple sources of funds, see Roger Geiger, *To Advance Knowledge: The Growth of American Research Universities, 1900–1940* (New York: Oxford University Press, 1986).
2. See, for example, *Pheromenes*, published by the Institute of Ecology, Academy of Sciences, Lithuania.
3. The census classifies 52.5 percent of the population as Latvian-speaking, compared with 34 percent Russian-speaking.
4. They are: (1) computer science; (2) mechanics, machine engineering, and energetics; (3) physics, mathematics, and astronomy; (4) technology of chemistry materials, chemistry, and pharmacy; (5) biology, ecology, geography, and geology; (6) molecular biology, microbiology, and biotechnology; (7) medical sciences; (8) agricultural sciences; (9) history; (10) linguistics, history of literature, and arts; (11) philosophy, sociology, psychology, and pedagogics; (12) economics and law; (13) scientology; and (14) wood sciences. See *Latvian Council of Sciences* (Riga: Government of Latvia, 1992).
5. *Annual Report*, Latvian Academy of Sciences (Riga: 1992), p. 3.
6. Social security is calculated at the self-employed rate—that is, 37 percent of base salary or 0.19 percent of the total.
7. Danish Research Council, Copenhagen, December 1992.
8. This summary cannot do justice to the wealth of information, care, and judgment found in this valuable report. In many cases, the generalities are far less important than the comments on particular sectors, institutions, and projects.
9. The funds are Road Fund; Agriculture Fund; Forestry Fund; Environment Fund; Scientific Research Fund; Information Fund; Social Security Fund; Medical Insurance Fund; and Innovation Fund (see World Bank Report, p. 294).
10. These were, together with the number of institutions: Astronomy and Physics (6); Informatics and Technical Sciences (3); Biology, Geology, and Chemistry (8); Humanities and the Social Sciences (5); and the central administration (4), including the library; see *Annual Report* (1991).
11. Estonian Academy of Sciences, *Annual Report* (1992).
12. See O. Aarna, "Higher Education and Research in Estonia," mimeo (Tallinn: 1992) for interesting remarks on compartmentalism and psychological compartmentalism, in particular p. 6ff.
13. Helsinki University Services, Ltd., *Estonia's Higher Education Development Needs: Identification Study* (Helsinki: 1993). The full list is found on page 2: "1) Lack of higher education policy and legislation at the national level; 2) Lack of national level higher education budgeting system; 3) Inefficient University administrative structure; 4) Lack of information network and insufficient access to different sources; 5) Poor technology transfer system; 6) Inadequate integration of research and education; 7) Inadequate system to gather information on labour

market needs; 8) Inadequate continuing education system; 9) Unrelevant diploma and degree programs; 10) Unqualified teaching staff."

14. O. Aarna, *op. cit.*, p. 4.

15. Described in their *Annual Report* (1992) as follows: "Freedom of action in making decisions concerning scientific work according to their research profile, every right to regulate all their activities (including the use of financial resources and administering of their property) at their own discretion. It reduced central governing functions of the Academy to a minimum, with centralized guidance over capital building also coming factually to an end in 1992."

16. Separate reports were written for natural sciences, which have been published, and for the social sciences and the humanities, which are still in draft form.

17. O. Aarna, *op. cit.*, p. 5.

18. Higher Education and Research Committee, Legislative Reform Program, Council of Europe, *Baltic States, Cooperation for Quality Assurance in Higher Education*, Report of the Regional Advisory Mission, Riga, October 24–26, 1994 (Strasbourg: Council of Europe, February 14, 1995).

Chapter Five
Poland
Higher Education and Science—A Survey

Zbigniew M. Fallenbuchl

Development before 1989: A Historical Outline

The Soviet political and economic system was transplanted to Poland at the end of the 1940s; higher education and the scientific and technological infrastructure were reorganized and adjusted to the requirements of the Soviet-type state and command economy. Soviet advisers were present in all ministries and governmental agencies; under their supervision, the leaders of the Polish United Workers Party and Soviet-trained specialists, who usually had only a rudimentary preparation, effected some far-reaching changes. Prewar employees were brushed aside or were forced to operate within the system irrespective of their own views. At the same time, "many new, so-called Marxist economists, philosophers and other specialists were inducted into higher places of learning."[1]

In accordance with the Soviet model, the entire field was divided into three component parts: (1) universities and specialized schools of higher learning, responsible for education under the strict supervision of the Ministry of Education;[2] (2) the Academy of Sciences, with its research institutes responsible for basic and applied research referred to it by ministries responsible for various sectors of the economy, financed from the state budget and administered by the scientific secretary, who held the rank of cabinet minister; and (3) specialized research institutes and research and development (R&D) centers in various sectors of the economy under the supervision of the appropriate "branch" ministries and financed mainly from ministry budgets. This framework existed, with only some minor modifications, until the end of communist rule in 1989.

Poland has a strong university tradition and some old universities.[3] Also, during the interwar years there was a very dynamic university life. It was impossible to ignore this inheritance completely and, as a result, the Soviet model of higher education was modified when it was introduced in Poland. For example, the degree of doctor, instead of the Soviet "candidate," was retained, as was the way it was awarded. Older universities retained their names, and the newly established took the names of some great Polish scientists instead of communist leaders, as happened in other countries of the Soviet bloc. Further, the separation of teaching from research was never as complete as in the pure Soviet system. The academic personnel of universities and other schools of higher learning continued to be involved, to a certain extent, in research, and members of the academy and ministerial research institutes often had teaching appointments.[4]

Poland did share with other Soviet-type economies the administrative control of a heavy bureaucratic structure which the central planning mechanism believed was necessary for the development of science and technology; the system was organized, administrated, and planned in the same way as the rest of the economy. Decisions were taken at different levels: Some were made by the Party Central Committee, the Council of Ministers, and the Central Planning Commission; others were determined by the Committee for Science and Technology (established by the Council of Ministers in 1985 and composed of nominated members chaired by a deputy prime minister in charge of science and technology), or by branch ministries responsible for various sectors of the economy (agriculture, coal mining, iron and steel, textiles, and so on). Researchers with a given project had to deal with these numerous offices to ensure that it was covered by the overall plan, and often had to modify details for this purpose. Researchers had to apply for both funds and specific equipment and materials in order to obtain import permits and the use of foreign currencies. All this was time consuming and created inflexibility that resulted in unofficial or illegal activities and corruption.[5] The overcentralized and heavily bureaucratic model of the management of science and technology choked the development of science and technology, discouraged initiatives, strengthened the dominant position of bureaucrats, and reduced the possibility of direct cooperation between researchers from different institutes within and without the country.[6]

Higher Education before 1989: Number of Schools, Students, and Graduates

Polish higher-education policy emphasized specialized schools and academies at the expense of universities. It has been suggested that "this diversification of schools was convenient for the communist government which employed the principle: *divide et impera* and tried to counterbalance strong centers of intellectual dissent by small servile institutions."[7] This was not the only reason. The priority given to the professional schools reflected also the ideological biases of the Stalinist communist party, and the needs of the rapid industrialization drive, which favoured a narrow, practical specialization at the expense of well-rounded general knowledge.

A number of specialized schools were established during the regime's early years, while the number of universities and higher schools of technology were reduced. Table 5.1 shows the changes, by type of institution, over time. The changes shown in Table 5.1 indicate a considerable degree of organizational instability up to 1989, with a consequent adverse impact on the process of education.

Both the instability and changes in the relative position of universities and higher schools of technology among all postsecondary institutions are reflected in the annual enrollment of students and in the number of graduates leaving school in a given year (see Table 5.2). The instability reflected in these numbers is a product of the political situation and economic strategy.

The 1950s was the period of the first rapid industrialization drive, based on import substitution, with priority given to heavy industry, the compulsory collectivization of agriculture, and the imposition of central economic planning and administration.[8] The total number of students in postsecondary institutions increased from 125,100 to 157,500. The educational system was reorganized to meet "the needs of building socialism and educating personnel to run the nationalized and centralized economy."[9] In particular there was a large increase of students (and graduates) from institutes of technology and agricultural academies. These numbers reflect the party's and planners' allocation preferences.

In the five-year period 1950–51 to 1955–56, the higher schools of technology graduated almost half of the total graduates from all postsecondary institutions. Engineers played a very important role

Table 5.1 Institutions of Higher Learning

Type of School	Year 1950/51	1955/56	1960/61	1970/71	1980/81	1985/86	1989/90	1990/91	1991/92	1992/93	1993/94
Universities	9	8	8	10	10	11	11	11	11	11	11
Higher schools of technology	20	15	15	18	18	18	18	18[30]	30	30	30
Agricultural academies	2	7	7	7	9	9	9	9	9	9	9
Economic academies	11	9	9	5	6	6	6	5	9	14	26
Higher pedagogical schools	7	6	4	9	11	10	10	10	10	11	13
Medical academies	11	10	10	10	10	11	11	12	12	12	11*
Higher schools of marine studies	—	—	—	2	2	2	2	2	2	3	3
Academies of physical education	4	4	4	6	6	6	6	6	6	6	6
Higher schools of fine and performing arts	19	17	16	16	17	17	17	17	17	18	19
Theological academies	—	2	2	2	2	2	7	7	8	8	8
Other	—	—	—	—	—	—	—	2	2	2	4
Total	83	78	75	85	91	92	97	96 [112]	117	124	140

Source: Glowny Urzad Statystyczny, *Rocznik Statystyczny* [Statistical Yearbook] (Warsaw, 1967), p. 457; 1973, p. 488; 1980, p. 408; 1989, p. 459; 1990, p. 473; 1992, p. 421; 1993, p. 439; 1994, p. 486.

Note: Numbers in brackets starting in 1990/91 include academic schools under the jurisdiction of the ministries of National Defense and Internal Affairs. After that date they are included in the totals. *Since Oct. 1, 1993, Cracow Medical Academy became Faculty of Medicine of the Jagiellonian University in Cracow.

in the newly created socialist command economy. They were needed as technical personnel, managers of state enterprises, bureaucrats and planners in the ministries and in various central agencies. A similar role was played in collectivized agriculture by the graduates of agricultural academies, enrollment in which increased from 3,400 to 15,900 students. There were also substantial increases in the number of students enrolled in the higher pedagogical schools and in economic academies, which supplied not only specialists in their respective fields but also large numbers of bureaucrats. In this five-year period the number of students enrolled at the universities declined from 33 percent to 16 percent of all students enrolled in postsecondary education; the number of university graduates declined from 24 percent to 18 percent of all graduates of institutions of higher learning.

In 1956, following the workers' riots in Paznan, Wieslaw Gomulka, who had been imprisoned by the previous leadership, became the first secretary of the Communist Party. There was a considerable dose of political liberalization, decollectivization of agriculture, some decentralization of the economy, a temporary change in investment policy, and a reduction in the pace of industrialization.[10] During the second half of the 1950s the proportions between special institutions (higher schools of technology and agricultural academies) and universities reversed themselves, reflecting a greater willingness to take the personal preferences of students into consideration. The number of students registering in economic academies continued to grow, as the decentralization of economic administration was expected to increase the role of economists in various sectors of the economy and they were, to some extent, substituted for engineers in management and bureaucracy.

The 1960s saw a new industrialization drive; by 1970 the number of students in higher schools of technology increased from 53,700 to 124,900, representing 38 percent of all postsecondary students in 1970. The year 1970 saw the introduction of Edward Gierek's "new development strategy" based on investment aimed at large-scale reconstruction and modernization of the economy using imported Western technology financed by substantial credits.[11] So long as Western credits were available, a rapid expansion of the economy was achieved. The total number of both students and graduates continued

Table 5.2 Number of Students [S] and Graduates [G] in an Academic Year in Thousands

Type of School		1950/51	1955/56	1960/61	1970/71	1980/81	1985/86	1989/90	1990/91	1991/92	1992/93	1993/94
		Year										
Universities	S	41.0	24.6	39.0	97.5	131.2	116.9	133.0	141.1	153.9	180.3	213.5
	G	5.3	3.9	4.5	14.5	21.0	18.3	17.9	19.3	19.9	22.5	n.a.
(as percentage of total)	S	[33]	[17]	[24]	[29]	[29]	[34]	[35]	[35]	[36]	[36]	[37]
	G	[24]	[18]	[20]	[31]	[28]	[31]	[33]	[33]	[32]	[35]	—
Higher schools of technology	S	34.6	63.6	53.7	124.9	127.6	71.9	72.0	84.0	93.5	109.2	130.6
	G	4.8	10.6	7.8	16.4	19.6	15.2	8.2	11.7	11.5	10.6	n.a.
(as percentage of total)	S	[28]	[40]	[32]	[38]	[28]	[21]	[19]	[21]	[22]	[23]	[22]
	G	[22]	[49]	[35]	[35]	[26]	[25]	[17]	[20]	[20]	[17]	—
Agricultural academies	S	3.4	15.9	13.9	33.5	61.0	35.5	35.9	36.4	36.2	40.9	46.7
	G	2.3	2.5	2.6	4.8	10.0	6.8	5.8	6.0	5.7	5.9	n.a.
Economic academies	S	14.6	16.2	21.4	25.0	34.1	21.5	23.6	24.0	28.5	37.9	53.1
	G	5.2	1.6	2.5	4.3	5.6	4.8	2.7	3.1	3.7	4.3	n.a.
Higher pedagogical schools	S	3.4	7.0	8.2	14.8	36.9	36.2	46.1	47.6	46.6	52.7	66.2
	G	0.3	0.3	0.7	1.8	7.2	5.7	6.9	8.4	8.4	9.2	n.a.
Medical academies	S	23.3	23.6	23.7	22.9	35.1	34.2	37.5	3.87	37.4	35.1	30.1*
	G	3.3	2.0	3.3	3.8	5.5	5.2	5.1	5.6	6.3	5.8	n.a.
Higher schools of marine studies	S	—	—	—	0.9	3.6	3.0	2.4	2.5	2.8	3.2	4.5
	G	—	—	—	—	0.6	0.6	0.5	0.3	0.4	0.5	n.a.
Academies of physical education	S	1.1	2.2	1.9	5.0	13.2	11.8	14.1	14.6	13.9	15.4	16.0
	G	0.4	0.3	0.4	0.8	2.6	1.7	2.0	2.4	2.6	2.6	n.a.

continued next page

Higher schools of fine and performing arts	S	3.7	3.7	5.2	7.7	7.5	8.0	8.2	8.3	8.5	9.0
	G	0.2	0.6	0.9	1.3	1.3	1.1	1.3	1.3	1.2	n.a.
Theological academies	S	—	0.5	1.1	1.8	2.2	5.8	6.7	7.1	8.5	9.0
	G	—	0.0	0.1	0.2	0.3	1.0	0.9	0.9	1.0	n.a.
Total	S	125.1	157.5	330.8	453.7*	340.7	378.4	403.8	428.2	495.7	584.0
	G	21.7	21.7	47.2	74.0	59.9	50.0	59.0	61.4	64.2	n.a.

Source: See Table 1.

Notes: 1. Starting in 1970/71, the number of graduates at the end of the academic year; 2. Starting in 1990/91, the numbers include students and graduates of academic schools under the jurisdiction of the ministries of National Defense and Internal Affairs.

*With special upgrading courses for teachers.

to grow, reaching their highest point at the beginning of the 1980s.

The division of students and graduates between types of institutions does not fully correspond to a division of the major academic disciplines. For example, economics could be studied at economic academies as well as several universities. The number of graduating students according to the major groups of disciplines is presented in Table 5.3. In 1980 the number of graduates reached 83,955. About 30 percent, or 24,904, graduated in technological subjects. Humanities, a most heterogeneous group that covers a wide range of subjects, represented the second largest group with 15,921, or 19 percent of the total. Economics graduates ranked third (12 percent), followed by natural sciences and mathematics (9.7 percent), legal and administrative studies, and agricultural studies. This composition reflects the priorities of the planned socialist economy.

In 1979 the economy entered into a deep economic crisis, directly related to arbitrary import cuts imposed by central planners to reduce the balance-of-payments disequilibrium.[12] The wave of strikes, the establishment of Solidarity, and the third enforced change in the communist leadership were followed by the imposition of martial law by General Wojciech Jaruzelski and the suppression of Solidarity.[13] Inconsistent and halfhearted economic reforms created a systemic vacuum—the old system had ceased to function and a new one was unable to replace it. The regime collapsed in 1989 as a result of prolonged economic stagnation, growing administrative disarray, lack of cooperation from the population, and a new wave of strikes.[14]

The last decade of communist rule had a truly devastating effect on postsecondary education. During the 1980s the total number of graduates drastically declined; in 1989, for example, there were 50,039 graduates, or 33,916 less than in 1980, with the greatest fall in technological disciplines (14,492), economics (5,270), legal and administrative studies (4,748), agricultural studies (3,428), and mathematics and natural sciences (3,154). The smallest declines were reported in medical sciences (385), fine and performing arts (525), physical education (893), and humanities (1,497). The only increase took place in the number of theology graduates (767) (see Table 5.3). At the end of the 1980s, the total number of students increased again; however, because of lags, the number of graduates decreased further.

Table 5.3 Graduates According to the Groups of Disciplines

Disciplines	Year 1937	1948	1950	1960	1970	1980	1989	1990	1991	1992	1993
Technological	667	747	2,816	7,877	16,883	24,904	10,412	10,857	11,980	11,433	11,521
Agricultural	497	540	1,133	1,671	3,554	7,210	3,782	3,544	3,732	3,906	3,947
Legal and administrative	2,479*	1,731*	5,744*	306	1,273	7,456	2,708	3,015	3,152	3,594	3,564
Economic	—	—	—	2,358	7,150	9,713	4,443	5,388	5,791	7,086	8,497
Humanities	990	537	1,391	1,570	5,630	15,921	14,424	15,828	17,181	16,877	18,680
Mathematics and natural sciences	453	308	815	1,710	4,665	8,162	5,008	4,968	5,259	5,587	5,986
Medical	974**	760**	2,520**	3,148	3,873	5,469	5,084	5,418	5,774	6,277	6,388
Physical education	—	—	—	526	620	2,998	2,105	2,341	2,556	2,762	2,817
Fine and performing arts	54	132	137	522	908	1,429	904	871	925	1,068	1,081
Theological	n.a.	—	—	—	—	402	1,169	1,004	1,086	1,055	1,131
Total	6,114	4,755	14,556	20,535	47,117	83,955	50,039	56,078	59,046	61,424	64,201

Source: G.U.S., *Rocznik Statystyczny* (Warsaw, 1967), p. 458; 1973, p. 493; 1990, pp. 476–77; 1992, pp. 424–25; 1993, p. 443; 1994, pp. 492.

Note: Starting in 1990, the numbers of technological and medical graduates include those who graduated from academies under the jurisdiction of the ministries of National Defense and Internal Affairs.

* Together with economics.

** Together with physical education.

The Quality of Higher Education

Until 1989, all institutions of higher learning were under the control of the Polish United Workers Party. The office of the party secretary of the local branch was usually close to that of the rector. Party approvals had to be obtained for all organizational changes, personnel policy including academic promotions, and even the distribution of various material benefits, student selection, and various campus activities.[15] Widespread political discrimination was particularly damaging to educational quality. A group of distinguished Polish academics described this situation, which continued during the 1980s, as follows:

Within the framework of personnel policy, party members' advancement was clearly accelerated, and various important functions . . . were assigned to them. Particular attention was paid to sending party members on excursions to Western countries, in order to profit from stipendiums and cooperation with foreign schools. On the other side, advancement opportunities for employees representing independent or opposition positions were limited and constrained.[16]

Study programs were prescribed by the ministry and were expected to be followed fairly rigorously. They all included ideological subjects, such as "Political Economy of Socialism," which were obligatory for all students irrespective of their field of study. Further, there were obligatory military preparedness courses as well as obligatory study of the Russian language. Obligatory courses added to the already overloaded programs of studies. Although the programs appeared narrowly specialized, they typically included relatively unimportant technical aspects of given fields while often missing an in-depth study of separate subjects. The ministry ordained textbooks, censorship restricted independent opinions, and access to foreign publications was limited, in some cases available only to authorized persons.[17] Large classes without any discussion, extensive use of relatively junior staff for teaching, and a reliance on hastily administrated oral examinations were typical features. Stress was placed on familiarity with lecture materials rather than independent studies and a capacity for analytical thinking.

Mathematics and natural sciences, technological subjects, medical science, and scientific aspects of agricultural studies were less affected by ideology. In contrast, humanities, particularly social sciences, were heavily biased.[18] This was especially true in the case of economics, which had a very specific program and heavily politicized the content of all courses. Graduates in this field lacked basic analytical tools of the discipline,[19] and "even including statistics and accountancy, were saturated with vulgar Marxist-Leninist doctrine."[20]

All information about Western countries and their economies was presented in a largely biased way. . . . The whole program of economic studies was subject to one key idea: to present the Soviet Union as a country of the most advanced technology, the best economy and the most developed welfare, and as the best example to follow. All possible efforts were made to prove the superiority of socialism over capitalism.[21]

"Branch specializations," corresponding to the groups of industries into which the central plans were divided, were introduced to train personnel for particular sectors of the command economy. They included courses on the economics of mining, engineering, the chemical, textile, and construction industries, and so forth, which described the organizational, administrative, and planning details together with some information about the relevant technologies and materials. But they provided no economic analysis. Even foreign trade specialization was a narrow training of administrative personnel that stressed organization and technique rather than the theory of international economic relations and an understanding of the world economy.

After the period of rigid political control and ideological indoctrination that was characteristic of the early 1950s, academic life became more liberal after 1956; this lasted until 1968. Older universities and other schools of higher learning were able to achieve some autonomy in less important matters, and their rectors, although all *nomenklatura* appointments, were sometimes able to make independent decisions on academic matters. Lublin, the only independent Catholic university, operated in a less restricted way.

There was some improvement in the quality of education, for example in economics, and between 1956 and 1968 some universities and academies were able to modify the way various courses were taught. However, as a result of student protests and riots in 1968, an anti-intellectual and anti-Semitic campaign "arranged and conducted by the Communist Party"[22] led to widespread purges of academic personnel. More rigorous party supervision was reintroduced and the institutions of higher learning lost whatever autonomy they had been able to develop.

There was a revival of independent academic thinking and activities during the short "Solidarity period" after August 1980. Various reform proposals were discussed and new course contents were introduced. However, the intellectual ferment at the institutions of higher learning ended with the introduction of martial law and the forced dissolution of Solidarity in December 1981. There was a new purge of academic personnel and the suppression of students' independent activities. A relatively liberal law on higher education, prepared during the short-lived "Solidarity period" as a compromise, was introduced in 1982, only to be amended in 1985. The remnants of any autonomy or internal democracy of institutions of higher learning were eliminated. Party control and discrimination against independent-thinking scholars increased.

A change in the financing of research, which took place in 1985, was heralded as a major reorganization and improvement in the administration and planning of the development of science and technology. In practice the modified system of planning, which was composed of a few central programs, sectoral programs, and plans for industrial enterprises, postsecondary schools, and research organizations,[23] further restricted freedom of scientific inquiry. Researchers were able to obtain financing for their projects only by linking them with one of a relatively limited number of selected programs. The political control over research did not decline but, on the contrary, increased. The climate became strongly unfavorable for the quality of education and research, as the following description well illustrates:

In the 1980s, especially after the launching of the system of financing of science in the form of the so-called government and ministerial "problems" (which most often indicated the financing of some chosen scholars) the functioning of coteries and groups of interests behind the official academic structures grew stronger. The criteria of evaluation of research and teaching were further decreased in favour of the protection of "the university's own people." There was virtually no sound competition in which qualifications and abilities are decisive; the staff was not em-

*ployed by way of competition (though it was a formal requirement); the
selection of young researchers and teachers was not based on evaluation
of their qualifications.*[24]

Natural sciences too were adversely affected by political interference. However, the principal obstacle came as a result of insufficient financing and very limited links with the West. This situation,
never satisfactory, drastically deteriorated after the introduction of
martial law (1981), and Polish research and teaching became increasingly isolated. Compared with social science graduates, natural
science graduates were reasonably well prepared, although the programs had major problems.[25]

The various fluctuations of party interference with postsecondary
education in the extent of the nonideologically determined content
of various courses, in the ability to establish and maintain contacts
with Western scholars, and in the selection of teaching personnel
and students (which often was "negative"—that is, weaker candidates who had little chance to be selected on merit looked for party
support and were able to achieve priority) were responsible for a
very uneven quality of teaching and research personnel in the institutions of higher learning and in the graduates of these schools. This
situation has been inherited by the postcommunist governments and
it will take a long time to overcome.

The Polish Academy of Sciences

The Polish Academy of Sciences was established in 1951. It evolved
with some "particular features, above all in functioning of the Academy in scientific life, in its relationship to the government and the
party, and last but not least in lacking the heavy military and industrial tasks known elsewhere."[26]

The academy combines three functions. First, it is a corporation of distinguished scholars who themselves elect new members
when openings occur. Although all appointments under the communist regime were subject to political approval, the academy retained, to a certain extent, the character of experts chosen by experts.[27] The situation differed from one period to another.

The election of the president of the academy was always subject
to strong political pressure and the scientific secretary had to be a

member of the communist party. He was appointed by the government with the rank of a cabinet minister. He nominated directors of the institutes and was responsible for the scientific operations of the academy, administration, and political conformity of its members. In the last year of the communist regime, party control was further increased when the scientific secretary, as the real power within the academy, became an ex-officio vice president.

The academy's membership is composed of full members, corresponding members, and foreign members. During the early years of the regime there were also titular members who had recognition but did not participate in the work of the academy. As Table 5.4 shows, the membership increased from 148 members, including 107 full and corresponding members in 1952 to 464 members, including 320 full and corresponding members in 1989. There was an increase in the number of foreign members during the 1960s and 1970s and again at the end of the 1980s and the beginning of the 1990s. Under the noncommunist governments the total number increased to 552 in 1991 and 505 in 1992 mainly because of a significant increase in foreign members.

Table 5.4 Polish Academy of Sciences: Members

Members	Year									
	1952	1960	1970	1980	1985	1989	1990	1991	1992	1993
Full members	34	72	131	176	163	194	186	205	193	186
Correspondents	73	115	119	137	140	126	124	134	133	130
Full and corresponding	107	187	250	313	303	320	310	339	326	316
Titular members	41	4	1	—	—	—	—	—	—	—
Foreign members	—	7	76	119	116	144	141	183	179	174
Total	148	246	327	432	419	464	451	522	505	490

Source: G.U.S., *Rocznik Statystyczny* [Statistical Yearbook] (Warsaw: 1971), p. 513; 1989, p. 435; 1990, p. 452; 1993, p. 418; 1994, p. 465.

The composition of membership of the academy strongly favors natural sciences and engineering. In 1989 the largest group, 142 scholars, or 31 percent of all members, were from mathematics, physics, chemistry, geology, and geography. Together with 54 biologists (12 percent), 41 percent of the members were from the natural sciences. Engineering scholars constituted 19 percent, medical scientists 12 percent, and agricultural specialists 11 percent. The total for the social

Table 5.5 The Polish Academy of Sciences: Members According to Groups of Disciplines

Groups of Disciplines

	1952	1960	1970	1980	1985	1989	1990	1991	1992	1993
Biology	30	27	34	46	47	54	54	63	61	58
Mathematics, physics, chemistry, geology, geography	25	59	88	132	125	142	138	161	157	154
Engineering	21	47	71	83	77	87	83	97	96	91
Medical	—	25	38	49	47	55	55	61	61	60
Agricultural	—	23	36	48	47	50	49	53	50	48
Social sciences, law, humanities	31	65	60	74	76	76	72	87	80	79

Source: See Table 5.4.

sciences, law, and humanities amounted to only 16 percent. These proportions were not changed during the early 1990s; in 1993, these same three areas were still only 16 percent of the total (see Table 5.5).

The academy, as a corporation of distinguished scholars, has 104 scientific committees, which represent various disciplines and act as consultative bodies. Approximately four thousand committee members are elected from the research community, the majority from universities and other postsecondary schools, for three years.[28] This is a heavy structure which, because of its size alone, cannot be efficient. It reflects a typically bureaucratic mentality of empire-building, as the power and prestige of the administrators increase with the size of the units for which they are responsible. Administrative costs connected with maintaining that structure and financing meetings of all these committees and their activities absorb a big chunk of science and technology funds, thus reducing research funds.

The academy also functions as the main state research center. It is composed of about ninety units, including fifty-one institutes, twenty research units, a number of specialized research centers, laboratories, libraries, archives, and other supporting offices. With a total staff of about 10,600, including about 4,400 scientists, in 1991 the research units of the academy represented about 77 percent of the research units of the country and, with few exceptions, were engaged in basic research.[29]

Until 1989 the academy's research units suffered under the joint weight of the party and the heavy internal bureaucracy. Research was closely monitored by the centralized and bureaucratic offices of the academy's scientific-section secretaries; this structure created barriers adversely affecting research publications, international exchanges, and other scholarly activities. The social sciences and humanities suffered most from the party's "political and ideological tutelage over them."[30]

The academy's third function was to act as the main governmental agency for the allocation of financial resources obtained from the State Committee for Science and Technology to various research programs and scholarly societies. This allocation was based on governmental priorities and the academy's own evaluation of the projects, institutes, and participating researchers. Although the allocation process was supposed to be based on scientific grounds, in practice, both political and personal connections played a decisive role.[31]

The academy exercised considerable power, which it often used to give priority to its own research or that of favorite institutes or scholars. This system of financing "has wasted a tremendous amount of money"[32] and ossified the positions of older researchers, irrespective of their real performance, and those of the established structures, programs, and projects. It was heavily biased against bright young scholars and new initiatives.

A major reorganization of the academy is an important task for the postcommunist government and will have to overcome strong vested interests.

Sectoral Research Institutes and R&D Units

There were, until the economic reforms of the 1980s reduced their number, eleven so-called branch (that is, sectoral) economic ministries. There were separate ministries for energy generation, mining, metallurgy, engineering industry, heavy and agricultural machine building, chemical industry, light industry (textiles, clothing, leather goods), construction and building materials, food processing, agriculture, and the forestry and wood processing industry. They controlled their respective sectors with the help of middle administrative units, the so-called associations, which were, in effect, an extension of the ministry responsible for all enterprises producing a particular group of commodities.

This organizational structure eliminated competition and, as the profits or surplus above the planned level were transferred by the association or ministry to finance unprofitable enterprises, there was no incentive to adjust production to potential demand or to reduce costs. As such, there was no inducement to introduce technological progress. Rather, any experimentation with new products or new methods of production might have reduced the possibility of fulfilling the quantitative targets and obtaining bonuses for management and workers.[33] The demand by enterprises for new technology was weak and was usually introduced at the time of building new factories or during a major reconstruction/modernization of existing plants.

All "branch ministries" had specialized research institutes, central laboratories, and research and development (R&D) centers; moreover they controlled the R&D units and laboratories that were attached to enterprises and associations. The Ministry of Science, Higher Education, and Technology was responsible for research units. Some "functional ministries," such as Finance, Foreign Trade, or Foreign Affairs, also had specialized research institutes. In addition, several independent, specialized central offices reported directly to the Council of Ministers. They included the Higher Office of Mining, Central Geological Office, Central Inspectorate of Energy Generation, State Atomic Agency, Central Office of Geodesy and Cartography, Committee for Norms, Measures, and Quality, the Patent Office, and the Central Statistical Office (GUS). They all employed scientific personnel and had some scientific and technological functions. Some were very large institutions. For example, GUS had more than twice as many employees as the Ministry of Science, Higher Education, and Technology and all other central offices taken together; and the Higher Office of Mining employed a greater number than the Ministry of Mining itself.[34]

In 1970 Ministry research units contained 10,551 scientific personnel while their central laboratories employed 343 scientists. Together these two types of sectoral institutions, laboratories, and research units employed 27 percent of scientific personnel, as compared with 2,209 researchers employed by the Academy of Sciences and 25,643 scholars employed by the institutions of higher learning. In 1970 the sectoral research units received 33 percent of all funds, while sectoral R&D centers received 63 percent; by 1980 the share of sectoral research units increased to become 49 percent of all research funds while the share of the R&D units declined to 25 percent (see Table 5.6).

Table 5.6 Employment of Scientific Personnel

Institutions	Year											
	1970	1975	1980	1985	1986	1987	1988	1989	1990	1991	1992	1993
Research units of the Academy of Sciences	2,209	3,807	4,541	4,203	4,305	4,437	4,538	4,587	4,388	4,385	4,020	3,972
Research units of branch ministries	10,551	15,165	17,387	13,897	14,187	13,374	12,718	10,912	10,552	9,314	8,396	7,820
Research institutes	9,702	12,863	14,446	11,879	11,914	11,449	10,802	9,426	9,054	8,158	7,379	6,752
Central laboratories	343	593	365	284	281	271	276	187	208	148	131	161
R&D centers	–	1,459	2,304	1,427	1,694	1,478	1,448	1,153	1,062	795	717	748
Scientific support units	254	304	355	189	186	163	175	160	148	117	98	60
Information centers	–	34	31	21	21	12	17	16	15	12	9	–
Scientific centers	7	63	88	68	66	67	75	66	67	59	55	20
Archives	190	141	213	90	89	73	64	50	41	44	32	37
Universities and other post-secondary schools	25,643	40,377	46,999	46,143	46,155	45,972	47,742	50,524	50,048	51,385	50,688	50,861
Total	38,657	59,653	69,282	64,432	64,833	63,946	65,172	66,183	65,136	65,201	63,202	62,713

Source: G.U.S., *Rocznik statystyczny* (Warsaw, 1978), p. 349; 1980, p. 384; 1990, p. 451; 1993, p. 417; 1994, p. 464.

In 1975, the greatest number of research institutions were directed toward industry. There were 918 of them employing 1,235 senior researchers with the rank of professor, 7,307 other scientific personnel, 102,986 technical personnel, and 12,488 administrative personnel (see Table 5.7). The second largest number of research institutions was in agriculture (171) and the third in construction (155). In comparison with these three sectors and with the total number of institutes, the number of research institutes under the jurisdiction of the Ministry of Science, Higher Education, and Technology was a modest 133, employing 1,206 professors, 4,552 other scientific personnel, 9,439 technical personnel, and 3,194 administrative staff.

Table 5.7 Outlays on Development of Science and Technology (current prices in millions of zloty and shares of totals)

Institutions	Year					
	1970	1975	1980	1985	1988	1993
Scientific and technical research units	5.8	15.3	20.5	51.0	211.5	7,822
Share [percentage of total]	[37]	[45]	[55]	[61]	[60]	[74]
Research units of the academy of sciences	0.7	1.4	2.3	7.1	37.0	1,449
Share [percentage of total]	[4]	[4]	[6]	[9]	[11]	[14]
Research units of branch ministries	5.2	13.8	18.2	43.9	174.6	6,373
Share [percentage of total]	[33]	[41]	[49]	[52]	[49]	[60]
Scientific support units	0.1	0.6	0.4	1.1	6.1	22
Share [percentage of total]	[0.1]	[0.2]	[0.1]	[0.1]	[0.2]	[0.2]
R&D units	10.1	13.8	9.2	18.8	66.9	1,332
Share [percentage of total]	[63]	[41]	[25]	[22]	[19]	[13]
Postsecondary schools	—	4.9	7.6	13.9	74.3	2,730
[percentage of total]	—	[14]	[20]	[17]	[21]	[26]
Total outlays on science and technology	15.9	34.0	37.4	83.8	353.3	10,574

Source: G.U.S., *Rocznik Statystyczny* (Warsaw, 1971), p. 600; 1980, p. 458; 1989, p. 439. These statistics have been discontinued in the general statistical yearbooks and are available for 1993 in a specialized publication only. G.U.S., *Nauka i technika w 1993 r.* [Science and Technology in 1993] (Warsaw 1994), p. 21.

The sectoral allocation of research capacity was inefficient. There was a great deal of duplication and a loss of economies of scale. Equipment and researchers with similar or the same specializations were allocated to several units where they were not fully utilized. Contacts between institutes belonging to different ministries were minimal and cooperation in research difficult to arrange. Many institute directors indulged in empire building while status-seeking ministers often wished to have a research institute of a substantial size, even when its personnel and equipment were not fully utilized. Research projects had to be relevant to a given sector and often differed from the research interests and even qualifications of individual researchers.

The attempt to restructure and modernize the economy with the help of credits and a large-scale technology transfer from the West in the 1970s demonstrated the weakness of the scientific and technological infrastructure. Although there were several reasons why Gierek's "new development strategy" did not give the expected results, principal causes were an inability to absorb the imported new technology, to adjust it efficiently to local conditions, and a failure to diffuse it broadly.[35] In 1981 a reorganization took place in Poland when four new ministries were formed from the old nine and one ministry was split into two in order to separate the Office of Maritime Economy (a branch ministry) from the Ministry of Foreign Trade (a functional ministry). At the same time, the power and the staff of the older functional and the new combined sectoral ministries increased.

The reorganization led to a growth in the total employment in central administration.[36] In contrast, the employment of scientific personnel in the ministerial research institutes, laboratories, and R&D centers declined during the first half of the 1980s. By 1989 it had returned to its 1970 level (10,912 as compared with 10,551), representing 16 percent of the total national scientific personnel (see Table 5.8). The share of the sectoral research units in total funding for research increased from 49 percent (1980) to 52 percent (1985), returning to 49 percent in 1988. The share of R&D units in the total research funds continued to decline from 25 percent (1980) to 19 percent (1989) (see Table 5.6).

The relative decline in the sectoral research units has continued

Table 5.8 Scientific and Technological Institutions According to Economic Sectors (number and employment)

Sector	Year	No. of Units	Total Employment	Professors	Other Scientific Personnel	Technical Personnel	Administrative Personnel
Industry	1975	918	187,190	1,235	7,307	102,986	12,488
	1980	849	152,496	1,042	8,637	84,894	9,906
	1985	644	109,913	876	7,477	39,193	11,368
	1988	622	68,112*	755	3,014	41,036	11,114
Construction	1975	155	43,900	203	903	32,316	4,632
	1980	115	12,829	163	908	8,139	1,317
	1985	71	5,563	137	759	2,323	681
	1988	77	5,470*	146	751	3,365	881
Agriculture	1975	171	14,669	232	1,254	7,414	964
	1980	148	10,701	230	1,173	4,665	622
	1985	170	9,731	246	996	2,620	682
	1988	158	6,083*	246	1,093	3,044	915
Forestry	1975	13	723	35	133	330	77
	1980	6	714	23	180	266	79
	1985	6	600	22	143	136	51
	1988	6	419*	17	118	209	56
Transport and communications	1975	35	4,837	121	652	2,557	425
	1980	30	4,579	91	675	2,434	467
	1985	25	3,556	72	488	1,320	505
	1988	23	2,779*	70	405	1,205	529
Commerce	1975	6	309	14	37	115	44
	1980	7	995	27	161	470	134
	1985	3	331	7	81	81	65
	1988	3	271*	6	64	38	60
Urban and rural infrastructure	1975	3	1,691	64	316	869	240
	1980	3	367	1	–	205	56
	1985	3	284	1	–	108	52
	1988	6	580*	6	70	176	109
Science and technology**	1975	133	25,474	1,206	4,552	9,439	3,194
	1980	172	26,939	1,354	5,756	9,134	3,485
	1985	173	21,752	1,301	4,600	5,424	3,511
	1988	169	19,120*	1,381	4,530	5,432	3,407
Health and social welfare	1975	18	7,018	184	1,204	3,368	581
	1980	18	8,675	216	1,376	3,842	742
	1985	18	7,381	241	1,320	1,821	1,160
	1988	19	3,891*	243	1,171	1,024	882
Physical education, sports, recreation	1975	1	152	5	75	15	40
	1980	2	251	8	111	50	44
	1985	2	214	6	96	49	32
	1988	2	225	8	102	42	34
Other***	1975	15	1,850	63	436	884	211
	1980	39	5,064	164	1,085	2,630	538
	1985	27	3,366	150	894	915	565
	1988	26	2,478	106	646	813	474
Total	1975	1,467	287,661	3,357	16,794	160,278	22,857
	1980	1,389	223,610	3,319	20,062	116,129	17,390
	1985	1,142	162,691	2,988	16,466	53,990	18,672
	1988	1,111	109,428*	2,993	15,376	56,384	8,461

Source: G.U.S. *Rocznik Statystyczny* [Statistical Yearbook] (Warsaw, 1978), p. 350; 1989, p. 434. These statistics have been discontinued since 1990.
Notes: "Other Scientific Personnel" includes adjuncts, senior assistants, and assistants.
"Technical Personnel" includes technicians with postsecondary and secondary technical education.
*Without manual workers.
**Scientific and Technological Research Units under jurisdiction of the Ministry of Science, Higher Education and Technology (after 1985, the Ministry of Science and Higher Education).
***Public and Political Administration, Defense, Law and Order.

under the postcommunist governments. Their total number declined to 9,314 in 1991 and 7,820 in 1993 (see Table 5.6). Some sectoral research units have undoubtedly been very useful and they might be retained while others could be transferred to a central research center to ensure economies of scale, eliminate duplication and excessive separation of various fields, and encourage research that may be of a particularly great importance for the whole economy. In the market economy the administrative divisions between industries disappear and, with some exceptions, there is no reason why research should be organized in this way. Diversified industrial firms may be interested in cooperation with several research institutes specialized according to type of problem rather than type of industry.

The Research Infrastructure

In 1970 the academy had 2,209 scientists out of the total of 38,657, or 5.7 percent, and received only 4 percent of total research funds. Postsecondary education did not receive any special funds for research, and 0.1 percent was allocated for various research support units. The relative importance of the academy increased in the 1970s, declined in the early 1980s, and increased again at the end of the decade. It employed 4,541 researchers in 1980 and 4,587 in 1989. A larger decline took place in the early 1990s (see Table 5.6). Its funding increased to 11 percent of total outlays on research in 1989. There was also a rapid increase in the scientific personnel of postsecondary education that began to receive research funding during the 1970s, reaching 46,999, or 68 percent of the total available personnel, in 1980, and 50,524, or 76 percent of the total, in 1989. Its share of funding was already 14 percent in 1975 and increased to 21 percent by 1988 (see Tables 5.6 and 5.8).[37]

Research concentrated on applications rather than basic research. In 1970 there were 229 scientific and technological research units, and out of that total, 139 research units, or approximately 61 percent, were classified as engineering. They employed 59 percent of professors, 59 percent of the scientific personnel, and 68 percent of technical personnel (see Table 5.9).

This large applied-research infrastructure had a relatively small basic-research foundation, and these proportions could have had an adverse impact on the development of both science and technology.

Table 5.9 Scientific and Technological Research Units According to Major Groups of Disciplines

Groups of Disciplines	Research Units	1970	1980	1988	1990	1991	1992	1993
Natural sciences and mathematics*	Number	27	41	53	55	70	74	80
	Employment	7,274	13,372	15,592	12,913	17,172	14,820	13,883
Engineering	Number	139	256	224	205	207	168	202
	Employment	51,922	104,210	68,376	54,647	44,399	35,062	29,065
Medical	Number	18	22	26	23	24	26	26
	Employment	6,008	9,473	5,730	4,124	4,237	3,917	3,647
Agricultural	Number	19	17	19	18	22	25	35
	Employment	4,993	6,079	6,589	5,755	6,235	7,359	8,082
Social sciences, law, humanities	Number	26	57	52	36	48	40	49
	Employment	2,662	8,345	6,004	5,311	4,743	3,979	3,673
Total	Number	229	393	374	337	371	333	392
	Employment	75,639	141,479	102,291	82,750	70,461	65,137	58,350

Source: G.U.S., *Rocznik Statystyczny* [Statistical Yearbook] (Warsaw, 1978), p. 351; 1989, p. 433; 1993, p. 417.
*Mathematics, physics, chemistry, biology, geology, and geography.

High technology is closely connected with modern physics, chemistry, and biology, and in 1970 it consisted of only twenty-seven research units in the field of natural sciences and mathematics, employing 339 professors, 1,349 senior researchers, and 3,968 technicians.

The infrastructure of medical research was also relatively weak. In 1970 there were eighteen research units in this field, employing 162 professors, 1,111 researchers, and 3,371 technicians. The entire area of social sciences, law, and humanities was served in 1977 by twenty-six institutes with 309 professors, 759 researchers, and 799 technicians.

The weakness of the research infrastructure in natural sciences and mathematics, in medical science and in social sciences, law, and humanities was recognized and some expansion in these fields took place during the 1970s and 1980s (see Table 5.9). The number of institutes in mathematics and natural sciences increased from 27 (1970) to 41 (1980) to 53 (1988), and during the same period the medical research units increased from 18 to 22 to 26; in the social sciences, laws, and humanities, the number of research units increased from 26 (1970) to 57 (1980) and then declined to 52 (1988). The number of engineering research institutes also increased (139 in 1970, 256 in 1980) with, however, a reduced number of professors for whom researchers were probably substituted as the number of the latter increased from 6,171 in 1970 to 10,568. The number of engineering research units declined to 224 in 1988, the number of professors employed declined to 909, and the number of researchers declined to 6,879. In the early 1990s this trend has continued (see Table 5.9).

Engineering was allocated 59.4 percent of all research funds in 1970, increasing to 80.3 percent in 1975, when the transfer of Western technology was at its maximum; it subsequently declined to 58.9 percent (1988) (see Table 5.10). The share for natural sciences and mathematics dramatically declined from 22.1 percent to 7.3 percent in 1975 and started to grow again to reach 20.8 percent in 1988. This comparison suggests a possible maladjustment between an overexpanded applied technological research and the very severely underfunded base of fundamental research.

The share of funding for medical research also declined quite drastically from 7.8 percent in 1970 to 3.2 percent (1975), increasing, subsequently, to 8.2 percent (1988), when it was recognized that funding to engineering was now excessive. The share of agricultural research declined from 8.3 percent (1970) to 4.6 (1980) and then increased to 6.4 percent in 1985 and 7.1 percent in 1988.

The allocation of research funds under communism was directly related to the Soviet-type development strategy responsible for the creation of the inefficient industrial structure,[38] which now makes recovery so difficult. There was a considerable overexpansion of heavy industry, closely connected to engineering research, and an underdevelopment of high-technology industries. In the 1950s and 1960s, when priority was given to steel metallurgy, coal mining, and so-called "heavy chemistry" (for example, sulphuric and phosphoric acids or fertilizers), plastics, and other petrochemicals, electronics and computer research was neglected. Again in the 1970s, mechanical engineering was given priority and the construction industry and telecommunications neglected. And, most important, during the whole postwar period, agriculture was drastically neglected, despite its great importance to the economy, and there was a very significant deterioration of health services. The underfunding of research in agriculture must be recognized as a very serious mistake for a country where agriculture and the food processing industry provide a significant part of exports; where during the whole postwar period agriculture desperately needed modernization and expansion; and where its neglect contributed to the severity of the economic crisis at the end of the 1970s.[39]

Table 5.10 Allocation of Outlays on Science and Technology in Research Units According to Groups of Academic Disciplines

Academic Disciplines	Year 1970	1975	1980	1985	1988
Natural sciences and mathematics	1,293	1,107	2,265	7,683	44,007
[percentage of total]	[22.1]	[7.3]	[11.0]	[15.1]	[20.8]
Engineering	3,471	12,252	15,518	34,588	124,608
[percentage of total]	[59.4]	[80.3]	[75.6]	[67.8]	[58.9]
Medical	457	481	871	2,860	17,304
[percentage of total]	[7.8]	[3.2]	[4.2]	[5.6]	[8.2]
Agricultural	485	695	884	3,255	15,086
[percentage of total]	[8.3]	[4.6]	[4.3]	[6.4]	[7.1]
Social sciences, law, humanities	142	725	988	2,613	10,537
[percentage of total]	[2.4]	[4.8]	[4.8]	[5.1]	[5.0]
Total	5,848	15,263	20,525	50,999	211,542
[percentage of total]	[100.0]	[100.0]	[100.0]	[100.0]	[100.0]

Source: G.U.S., *Rocznik Statystyczny* (Warsaw, 1980), p. 391; 1989, p. 439. These statistics were discontinued after 1990.

The first decline in national product in 1979 demonstrates the seriousness of the economic crisis and was followed by a prolonged stagnation and further decline in 1989, forcing the regime to reduce outlays on science and postsecondary education. As noted, outlays on science never represented a large share of the total state budget expenditures, so its decline during the 1980s was particularly serious. Outlays on postsecondary education grew during the 1980s and reached 2.8 percent in 1989 (see Table 5.11).

As the result of this uneven development, which was heavily biased toward central planners' preferences, Poland constructed a considerable but skewed research infrastructure. During the whole postwar period, research suffered from the lack of adequate financing and relative isolation from international contacts outside the Soviet bloc. During the 1980s the situation deteriorated further and will now require greater modernization and restructuring of the scientific infrastructure to bring it to world levels and to parity with Western science.

Table 5.12 illustrates the present situation concerning current research outlays, employment and value of scientific equipment. There still is a clear bias in favor of natural sciences at the expense of social sciences and humanities and, within the former group, in favor of technical sciences at the expense of basic sciences. The table shows how limited are funds from foreign sources, which include

international financial institutions. It also shows the very high proportion of scientific equipment that has deteriorated and requires
replacement, not even taking into consideration the need for modernization.

Postcommunist Reforms

General Situation and Obstacles to Change   Since the communist regime collapsed, it has been recognized that a major restructuring of Polish higher education and scientific research is necessary.[40]
A new Higher Education Law was enacted in September of 1990,
mapping considerable reform of universities and other institutions
of higher learning. The State Committee for Scientific Research was
established by a law enacted in January of 1991 and the Committee
was formed in May of that year with the objective of providing leadership in transforming the entire scientific and technological infrastructure.

Postcommunist governments have recognized the importance
of higher education and science and technology in building a modern democratic state and creating an open-market economy. However, it should be recognized that only a little has been achieved in
this field since the collapse of the communist regime in 1989. The
reasons are (1) an unavoidable political instability and administrative confusion at the beginning of the gigantic systemic transformation in the country; (2) a serious long-run structural economic crisis
and macroeconomic instability inherited from the communist governments, and the difficulty and length of the economic stabilization and restructuring process;[41] (3) difficulty in achieving consensus on specific proposals concerning reform of science and
postsecondary education; and (4) the existence of psychological and
other barriers to change. In the long run, these psychological obstacles will likely be the most difficult to cope with.

The economic barriers to change are also important. Human
capital improvement and the acceleration of technological progress
are the necessary conditions for systemic transformation, restructuring and modernization of the economy, and sustained long-run
growth. However, the country is in a very serious economic crisis.
Although a recovery started in 1992 and has continued in 1993 and

Table 5.11 Expenditure from the State Budget on Science and Postsecondary Education [current prices, billions of zloty]

Expenditures	Year 1970	1980	1985	1986	1987	1988	1989	1990	1991	1992	1993
Total state budgetary expenditure	329	1,366	3,488	4,193	5,031	8,431	29,618	172,165	241,858	381,890	502,428
Expenditure on science	3	16	31	32	10	69	109	422	6,129	7,404	8,928
Share of total expenditure [percentage]	[0.9]	[1.2]	[0.90]	[0.77]	[0.20]	[0.81]	[0.34]	[0.25]	[2.5]	[1.9]	[1.8]
Expenditure on postsecondary education	5	19	74	88	110	206	836	6,212	6,650	10,110	12,696
Share of total expenditure [percentage]	[1.6]	[1.4]	[2.1]	[2.1]	[2.2]	[2.4]	[2.8]	[3.6]	[2.7]	[2.6]	[2.5]
Gross domestic product [billion zloty]	n.a.	n.a.	10,445	12,953	16,940	29,629	118,319	591,518	824,330	1,149,442	1,557,800
Outlays on science as percentage of GDP	–	–	[0.30]	[0.25]	[0.06]	[0.23]	[0.09]	[0.07]	[0.74]	[0.64]	[0.57]
Outlays on postsecondary education as percentage of GDP	–	–	[0.71]	[0.67]	[0.65]	[0.69]	[0.71]	[1.05]	[0.81]	[0.88]	[0.81]
Outlays on science and postsecondary education as percentage of GDP	–	–	[1.01]	[0.92]	[0.71]	[0.92]	[0.80]	[1.12]	[1.55]	[1.52]	[1.38]

Source: G.U.S., *Rocznik statystyczny* (Warsaw: 1976), pp. 495, 497; 1980, pp. 458, 459; 1985, pp. 94, 97; 1990, pp. 111, 140; 1993, pp. 130, 146; 1994, pp. 140, 156.
n.a.: Not available.

1994, the difficulties of balancing the state budget, relatively low personal incomes, and the limited size of an affluent private sector will continue for some time.

A drastic shortage of financial resources, caused by the necessarily restrictive monetary and fiscal stabilization policy, is a dominant obstacle to large-scale modifications in the short run. At the time when a major effort is being made to reduce domestic absorption (domestic expenditure on consumption, investment outlays, and public expenditures at all levels of the government) in order to curb inflation and to expand exports in order to secure some necessary imports, it is difficult to finance education and science, even though their importance is fully recognized. As Table 5.11 shows, in 1991 and 1992 the share of science in the state budget increased but that of postsecondary education declined. In 1992 only 0.65 percent of GDP was allocated to science and 0.88 percent to postsecondary education, and in 1993, 0.57 and 0.81 percent, respectively.

Higher Education The reform of postsecondary education has concentrated on: (1) establishing autonomy and internal democracy in the universities and other schools of higher learning (the Higher Education Law was adopted in September of 1990); (2) removing ideological and political bias from the contents of academic courses; and (3) revising study programs in various disciplines. The Marxist indoctrination and military preparedness courses have been removed and the teaching of Russian has ceased to be obligatory. In economics, narrow sectoral specializations have been replaced by macro- and microtheory, mathematics, and quantitative studies, and business management has become a distinct discipline. There is, however, a shortage of university teachers, especially in the social sciences, who can teach new courses, and not all current faculty are able to meet new course requirements.

There are proposals for a fundamental change in the nature and organization of postsecondary schools and a general modernization of academic programs to be designed and introduced by the schools themselves. The reformed Ministry of National Education will be expected to facilitate rather than prescribe. There are, however, voices calling for the Ministry of National Education to be more active than envisaged by the new legislation, because "the introduction of

Table 5.12 Outlays, Employment and Value of Scientific Equipment in Scientific and R & D Institutes in 1993 (European Classification of Research Fields)

Research Fields	Total Outlays (Million Zloty)	Foreign Funds (Million Zloty)	Total Employment	Scientific Personnel	Gross Value of Equipment (Million Zloty)	Percentage Requiring Replacement
Natural sciences	7,728,730	98,912	54,840	10,345	3,654,711	74.5
Mathematics and physics	357,333	1,514	1,894	803	265,248	88.9
Chemistry	696,110	3,568	4,300	780	443,533	72.2
Earth sciences	796,819	11,589	3,722	718	168,857	58.0
Biology	317,834	5,475	1,762	703	166,847	79.1
Forestry, agricultural and veterinary studies	1,027,894	33,314	10,925	1,893	327,495	66.4
Medical studies	764,712	4,609	3,050	1,371	354,535	60.8
Technical studies	3,746,006	38,070	29,078	4,035	1,917,889	78.0
Other natural sciences	22,022	773	109	42	10,307	83.7
Social sciences and humanities	569,163	3,339	3,578	1,447	43,203	60.3
Economics	146,329	1,273	880	257	1,830	20.4
Law	38,956	0	240	92	9,410	37.6
Philosophy and sociology	24,275	173	143	80	0	–
Linguistics and literature	47,400	210	353	203	2,676	45.7
Archeology and history	74,614	1,293	619	361	9,456	65.0
Fine arts	37,024	0	151	71	0	–
Theology	17,935	0	172	14	2,357	82.7
Physical culture	29,792	0	162	69	0	–
Other social sciences and humanities	152,838	390	858	300	17,474	73.3
Total	8,329,568	102,251	60,701	11,852	3,698,302	74.4

Source: G.U.S., *Nauka i technika w 1993* [Science and Technology in 1993] (Warsaw: 1994), pp. 80–81.

genuine changes in the existing universities demands today that certain mechanisms should be imposed . . . and this entails a temporary suspension of their autonomy."[42]

Although this latter argument is based on a plausible analysis of postsecondary education, it should be rejected. Just like democracy, which cannot be imposed from above but has to be learned by practice, postsecondary-education reform enforced from above will not be successful unless it is both well understood and accepted throughout the system. Its implementation requires real political leadership rather than administrative coercion.

A modified version will be more acceptable. It would grant a basic autonomy to schools but recommend that the Ministry on National Education "must conduct a decisive policy of supporting (especially financially) definite solutions and must discriminate visibly in resource allocation against the institutions and persons not involved in the process of change."[43] If conducted consistently, such a policy of steering reforms with the help of carefully differentiated financial grants may prove to be very effective.

Recommendations   It would be feasible at this stage to create a semiautonomous state system of postsecondary education. Tuition income would be expected to cover only part of the total expenses, and state grants therefore would be needed. If state funds are used, it is only logical that there be control over how they are spent. There should exist, however, a body which would represent the schools in their contacts with the government and which would protect them from direct interference by the relevant ministry.

Although the schools should have freedom to teach courses according to their own programs, some coordination of fields of study by the ministry, or preferably, by a committee selected by the schools themselves, is needed to eliminate unnecessary duplication, on the one hand, and to ensure, on the other, that all required fields are made available within the system as a whole. Grants might be used to ensure the desirable structure of the system; in order to achieve higher standards, an accreditation system should be established for every major academic discipline, preferably to be operated with the help of committees selected by the schools. A considerable waste of resources and time can be avoided by the use of foreign advisers familiar with the operation of Western systems. Although the establishment of private schools should be allowed and indeed encouraged, they also should be subject to appropriate accreditation procedures.

It may be necessary to reduce the number of independent specialized schools. Significant economies could be achieved by incorporating economic, pedagogical, physical education, and the performing and visual arts academies with universities operating in the same location. The amalgamation of administration would reduce overhead costs. Joint ownership of libraries and laboratories would help their necessary modernization and prudent expansion. Some courses could be given to students from different fields. Closer academic contacts of the staff of the academies with the scholars in the traditional university disciplines would be beneficial and stimulating to both sides and so create new opportunities for researchers.

Instead of a reduction in the number of specialized institutions, their number increased in the academic year 1992/93 and again in 1993/94 (see Table 5.1). This does not seem to be a rational policy, taking into consideration the shortage of fully qualified staff and funds and the inadequate level of libraries, laboratories, and other facilities. It would be more rational to close down a considerable number of schools and concentrate resources on the best of those remaining.

Some changes in the general structure of curricula will also be beneficial. The former rector of Warsaw University has suggested the following very useful changes in this respect:

1. *Present curricula aimed at producing narrow specialists should be replaced by curricula providing more general and basic knowledge (in order to enhance the graduates' general ability to learn and adapt).*
2. *Present rigid curricula which are, in fact, "timetables" for the students, should be replaced by a flexible credit system of courses enabling the students to shape their own profile.*
3. *The present necessity for the students to choose final specialization at entry to a university must be abandoned.*
4. *Credit transfer system must be introduced to enable student mobility within Poland and within Europe.*
5. *Accompanying the "massification of higher education," a degree below master's level should be introduced (e.g. bachelor's or associate's).*[44]

These proposals would improve the system and make it more flexible. They deserve to be carefully studied for broad implementation.

Modification of the Academy of Sciences Since the end of the communist regime, only relatively minor changes have taken place in the Academy of Sciences. Various proposals have been debated within both the research community and government. The future of the academy has not yet been decided. The proposals range from the demand to abolish the academy altogether to leaving its structure basically unchanged while reducing quite considerably the degree of internal centralization and the size of its administrative staff. Other proposals call for the academy to function as a corporation of distinguished scholars only, transferring its research institutes to a

new separate state research agency or to universities and other postsecondary schools as "university research centres." Still another solution would be the formation of semiprivate or regional organizations and "to entrust them with running selected research institutes" with "financial support from the state funds."[45]

The academy's top administration changed in 1990, a reorganization was commenced, and the administrative staff was reduced by 30 percent. The present president of the academy seems to advocate the retention of the two divisions with some modifications in both. In the academy as a corporation of distinguished scholars, a special group of "seniors" would be established to which all members seventy-five years or older would belong. They would not be counted in the statutory number of members, which is at present limited to 365. This would create the possibility of appointing a number of younger members, as well as those who could not be appointed previously because of their political views or their independence in representing new and unorthodox scientific ideas. It has also been suggested that the statutory maximum number of members could be increased.[46] This change would further accelerate the process of institutional cleansing and renewal.

There have been some structural changes with regard to the academy's research centers. Several social science institutes, particularly the most ideological and political, have been closed or transformed. Institutes have been given a considerable degree of autonomy, with the directors elected by scientific councils and administrative staff appointments made as the result of open competition. A council of institutes has been established as an elective body representing all scholars of the academy's institutes. It has not yet been decided if all or any institutes will remain attached to the academy.[47] It has, however, been suggested that the "Polish Academy of Sciences as a corporation does not seem to be too interested in reforming its anachronistic structure" and that "the efforts of a few reform-oriented PAS Members did not produce any visible results."[48]

Recommendations   It would be useful to retain the academy as a corporation of distinguished scholars able to express an independent expert view on policies concerning higher education and science; to evaluate the state of arts and sciences in the country; and to

discuss problems facing postsecondary education and scientific research. To perform this role the academy has to be completely independent from the government, with membership based on elections.

There is also need for a central research organization. The French *Centre National de la Recherche Scientifique* (CNRS) provides a good model for a country in which research will have to be financed mainly by the state. It is doubtful that affluent private individuals or foundations will provide sufficient funds to finance research for some time to come. The central research organization could take over the academy research institutes and some, but not all, sectoral research units to achieve economies of scale. It should be autonomous and insulated from direct pressures by the government and should not be involved in the allocation of research funds to postsecondary schools and other research institutions.

Sectoral Research Units The future of the large number of sectoral institutes is still unknown. It has been suggested "that in the present situation, the only reasonable science policy . . . is to identify those domains and institutions of basic and applied research which manifest the highest competence and performance, and those which have essential importance for the educational system and national culture."[49]

It appears that this approach has been adopted. The Committee for Scientific Research, established in 1991, is composed of representatives of the national and local governments as well as representatives elected by the research community. It consists of two divisions that deal separately with basic and applied research. Research grants are allocated on the principle of peer evaluation in accordance with a published formula; all research institutes and R&D units were invited to apply in July of 1991.[50]

The competition results were published in February of 1992 and created serious concern, especially from formerly high-priority institutes that were allocated a relatively low classification in the competition and, therefore, very limited funding. For example, the prestigious Institute of Finance, which has served as the research arm of the Ministry of Finance, obtained only a "C" classification. In a press interview, the high-profile director of the Institute strongly protested and accused the Committee of unfairness and a hidden agenda.[51] Irrespective of merit, these protests were unavoidable, as competitive grant allocation through peer evaluation was previously unknown in Poland.

In December of 1992 the government of Prime Minister J.K. Bielecki reaffirmed the principle of allocating grants by competition and peer evaluation, and it adopted the guideline that funding priority in basic research should be given to "those teams and institutes who have the best world standard achievements or are necessary for the maintenance of the necessary level of civilization for the society." For applied studies, financial support should be given to the research projects necessary "for the satisfactory functioning of the health services, environmental protection, agriculture, the economic and social infrastructure and will improve competitiveness of Polish exports." The government also announced that legislation regarding research institutes, together with a modification of the legislation on the Committee for Scientific Research, were to be submitted soon to the Parliament.[52] Because of the subsequent governmental changes, it is not clear whether the present government will proceed.

Recommendations A number of the sectoral institutes could be transferred to the central research center, if created, or to the academy institutes, if they are retained, or to the postsecondary schools. Some could be liquidated without a significant adverse effect on science and technology. Still others could be privatized and made dependent on self-financing. It may also be useful to retain a certain number of fully or partly government-financed research institutes to serve some particularly important sectors of the economy. Agriculture is an obvious such field, in which governmental research institutes exist in many Western countries. There may be other fields in which specialized governmental research institutes would be beneficial.

Industry's Interest in Technological Progress

In the past the lack of competition and other systemic features of the highly bureaucratic command economy made industrial enterprises insensitive to technological progress. They were always able to dispose of their products whatever their quality. There was no demand by the enterprises for the output of the science and technology sector, and a gap appeared between industry and the academy, sectoral R&D units, and research institutes of the institutions of higher education. There was no correlation between the amount spent on science and technology and industrial innovation. The science and technology sector was effectively separated from the rest of the economy.

Economic reform has not yet modified significantly the behavior of the state-owned industrial firms that still dominate industry, especially large-scale industry, and are likely to remain dominant for some time. Where the domestic and foreign demands for their products have declined, the enterprises must innovate to survive. The need to shift exports from the former Soviet markets to Western markets, in particular, requires the adoption of new products, modernization of processes, a reduction in costs of production, and the development of marketing skills.

So long as an enterprise may count on obtaining tax concessions and covert or even open subsidies in order to avoid reducing employment, or to avoid closing down its inefficient production, it will regard good relations with the authorities as more important than innovation. The demand for innovation will increase only when profit and loss really direct the activities of the industrial enterprises and they must depend on their own efforts, rather than help from the bureaucracy, for their survival.

Demand for innovation will remain limited and the separation of the science and technology sector from the rest of the economy will continue until the complete systemic transformation is effected. In the meantime there will be no correlation between expenditures on science and technology and actual innovation. As in the past, it almost does not make any difference what percentage of GNP is spent on science and technology. The completion of economic reform is a necessary condition for the technological progress on which economic growth depends.

Conclusions

As this survey indicates, after forty-five years of communist rule in Poland, fundamental changes are needed in higher education and science. This is necessary in order to accelerate a successful transformation from the Soviet-type totalitarian state and command economy to a democratic state and an open capitalist market economy. The process of transformation, restructuring, and modernization of higher education and science in Poland has only started; some difficult and often controversial decisions have yet to be made. The quality of the prepared proposals is of crucial importance. Technical and financial aid from international institutions and Western governments is extremely important for the success of this transformation. In the final analysis, the success of the process of transformation in this field, as in all other fields, will depend above all on the quality of political leadership. The changes undertaken must be carefully prepared and explained in order to secure the strongest possible support.

Notes

1. Janusz Beksiak, Ewa Chmielecka, Urszula Grzelonska, Aleksander Muller, and Jan Winiecki, *Higher Economic Education in Poland* (Warsaw: Program for Economic Studies in Central and Eastern Europe, July 1990), p. 5.
2. Two exceptions to Ministry of Education control were the Higher School of Social Sciences at the Central Committee of the party, which operated under the direct jurisdiction of that committee, and military and police (militia) schools under the jurisdiction of the Ministry of Defense and the Ministry of Internal Affairs, respectively.
3. The oldest university, the Jagiellonian University in Krakow, was established in the fourteenth century.
4. Alexander Gieysztor, "The Independence of Academic Institutions and Government: The Polish Academy of Sciences," in Antoni Kuklinski (ed.), *Transformation of Science in Poland* (Warsaw: State Committee for Scientific Research, 1991), p. 198.
5. For more details see Zbigniew M. Fallenbuchl, "An Overall Analysis of the Factors Impending Development and Progress of Civil Science," in Craig Sinclair (ed.), *The Status of Civil Science in Eastern Europe—Proceedings of the Symposium on Science in Eastern Europe, NATO Headquarters, Brussels, Belgium, September 1988* (Dordrecht: Kluwer Academic Publishers, 1989), pp. 353–58.
6. *Ibid.*
7. Andrej Kajetan Wroblewski, "The Future of Universities in Poland," in Kuklinski (ed.), *op. cit.*, p. 168.
8. Zbigniew M. Fallenbuchl, "The Communist Pattern of Industrialization," *Soviet Studies*, XXI (4), 1970, pp. 451–78.
9. Beksiak et al., *op. cit.*, p. 5.
10. See, for example, Zbigniew M. Fallenbuchl, "Investment Policy for Economic Development: Some Lessons of the Communist Experience," *Canadian Journal of Economics and Political Science*, XXIX (1), 1963, pp. 26–39.
11. The author has discussed this strategy in Zbigniew M. Fallenbuchl, "The Strategy of Development and Gierek's Economic Manoeuvre," *Canadian Slavonic Papers*, XV (1–2), 1973, reprinted in Adam Bromke and John W. Strong (eds.), *Gierek's Poland* (New York: Praeger, 1973), pp. 52–70, and in "The Polish Economy in the 1970s," in Joint Economic Committee, Congress of the United States, *East European Economies Post-Helsinki* (Washington, D.C.: U.S. Government Printing Office, 1977), pp. 816–64.
12. Zbigniew M. Fallenbuchl, "The Polish Economy at the Beginning of the 1980s," in Joint Economic Committee, Congress of the United States, *East European Economic Assessment* (Washington, D.C.: U.S. Government Printing Office,

1981) pp. 33–71, and "Poland's Economic Crisis," *Problems of Communism,* XXXI, (2), 1982, pp. 1–21.

13. Zbigniew M. Fallenbuchl, "The Economic Crisis in Poland and Prospects for Recovery," in Joint Economic Committee, Congress of the United States, *East European Economies: Slow Growth in the 1980s* (Washington, D.C.: U.S. Government Printing Office, 1986), pp. 359–98; "Poland: The Anatomy of Stagnation," in Joint Economic Committee, Congress of the United States, *Pressures for Reform in the East European Economies,* v.2 (Washington, D.C.: U.S. Government Printing Office, 1989), pp. 102–36.

14. Zbigniew M. Fallenbuchl, "Present State of the Economic Reform," in Paul Marer and Wlodzimierz Siwinski (eds.), *Creditworthiness and Reform in Poland* (Bloomington: Indiana University Press, 1988), pp. 115–30; "The Polish Economy in the Year 2000—Need and Outlook for Systemic Reforms, Recovery and Growth Strategy," *Carl Beck Papers in Russian and East European Studies,* No. 607, 1988, pp. 1–29; and "The 1986–1990 Five Year Plan: Strategy and Reform Dependency" in David M. Kemme (ed.), *Economic Reform in Poland: The Aftermath of Martial Law 1981–1988* (Greenwich, Conn.: JAI Press, 1991), pp. 17–37.

15. Barbara Markiewicz and Barbara Stanosz, "A Report on the University," in Kuklinski (ed.), *op. cit.,* p. 181.

16. Beksiak et al., *op. cit.,* p. 9.

17. The author's book, Zbigniew M. Fallenbuchl, *Polityka go spodarcza PRL* (Economic Policy of the People's Republic of Poland) (London: Odnowa, 1980), was placed in the library of the University of Lodz, together with other restricted materials, in a separate room; it was available only to authorized students and faculty members.

18. Markiewicz and Stanosz, *op. cit.,* p. 180.

19. Beksiak et al., *op. cit.,* p. 10.

20. Aleksander Müller, "Changes in the System of Higher Economic Education in Poland," in Kuklinski (ed.), *op. cit.,* p. 172.

21. *Ibid.,* pp. 172–73.

22. *Ibid.,* p. 174.

23. *Rocznik polityczny i gospodarczy 1986* (Political and Economic Yearbook) (Warsaw: P.W.N., 1987), p. 353.

24. Markiewicz and Stanosz, *op. cit.,* p. 181.

25. The author, together with a senior professor of physics, visited one of the best Polish universities, the University of Lodz, in 1978, in connection with the establishment of academic cooperation between that university and the University of Windsor in Canada. In order to select areas of mutual interests, all faculties, departments, institutes, and laboratories were jointly examined. The lack of even the most important Western scientific journals, the shortage of computers and other equipment, and the very poor state of the laboratories were clearly visible. The considerable dose of frustration caused by this situation among the faculty members was not difficult to discover.

26. Gieysztor, *op. cit.,* p. 201.

27. *Ibid.,* p. 202.

28. *Ibid.*

29. *Ibid.*

30. *Ibid.*

31. Markiewicz and Stanosz, *op. cit.,* p. 181.

32. Witold Karczewski, "Some Remarks on Science and Technology in Poland," in Kuklinski (ed.), *op. cit.,* pp. 15–16.

33. For an excellent study of this aspect of the functioning of the Soviet-type economies, see Joseph S. Berliner, *The Innovation Decision in Soviet Industry* (Cambridge, Mass.: MIT Press, 1976).

34. Zbigniew M. Fallenbuchl, "Science and Technology in Poland: The Infrastructure, Policies and Trends," in Herbert I. Fusfeld (ed.), *Framework for Interaction: Tech-*

nical Structures in Selected Countries (New York: Rensselaer Polytechnic Institute, 1988), pp. II. F. 1–37.

35. Zbigniew M. Fallenbuchl, *East-West Technology Transfer: Study of Poland 1971–1980,* (Paris: OECD, 1983).

36. Fallenbuchl, "Science and Technology," p. II. F. 8.

37. On the other hand, the sectoral research units and R&D units employed 10,551 researchers, or 27 percent of the total scientific personnel, and received 96 percent of all funds allocated for research.

38. Zbigniew M. Fallenbuchl, "Industrial Structure and the Intensive Pattern of Development in Poland," Jahrbuch der Writschaft Osteuropas, 4, 1973, pp. 233–54; "L'interaction de la stratégie de développement et du système économique, source de crises socio-économiques pérodiques en Pologne," *Revue d'études comparatives est-ouest,* 15 (1), 1984, pp. 113–30.

39. Zbigniew M. Fallenbuchl, "An Overview of the Role of Agriculture in the Polish Economic Crisis," in J.C. Brade and K.-E. Wädekin (eds.), *Social Agriculture in Transition* (Boulder: Westview Press, 1988), pp. 125–30.

40. See, for example, a report of the Society for the Advancement of Sciences and Arts on this subject, *Stanowisko w sprawie ustorju nauki* (The Position on the Organization of Science) (Warsaw: Towarzystwo Rozwoju Nauk, 1990).

41. Zbigniew M. Fallenbuchl, "Eastern Europe on the Road from Communism to Capitalism," *Canadian Business Law Journal,* 17 (2), 1990, pp. 238–45; "Poland: The Case for Cautious Optimism," *The World Today,* 47 (11), 1991, pp. 185–89; and "Transition from the Socialist Command Economy to the Capitalist Market Economy in Poland," in Günter Wagenlehner (ed.), *Von der Ost-West-Konfrontation zur Europäishen Friedensordung* (München, 1992), pp. 87–101.

42. Markiewicz and Stanosz, *op. cit.,* p. 183.

43. Beksiak et al., *op. cit.,* p. 23.

44. Wroblewski, *op. cit.,* pp. 168–69.

45. Andrzej Ziabicki, "The Role of the Academies of Sciences," in Kuklinski (ed.), *op. cit.,* p. 218.

46. Gieysztor, *op. cit.,* p. 204.

47. *Ibid.,* pp. 204, 205.

48. Ziabicki, *op. cit.,* p. 223.

49. Stefan Amsterdamski, "Difficulties in the Enforcement of the Reform," in Kuklinski, *op. cit.,* p. 295.

50. Malgorzata Sikorska, "Mistyfikacja?" (A Mystification?), *Zycie gospodarcze,* February 23, 1992, pp. 1, 4.

51. Tomasz Jerioranski, "Autor!! Rozmowa z dyrektorem Instytutu Finansow Prof. Grzegorzem W. Kolodko" (The Author!! An Interview with the Director of the Institute of Finance, Professor G.W. Kolodko), quoted in M. Sikorska, *Ibid.,* p. 4.

52. "Zalozenia polityki spoleczno-gospodarczej na 1992 r." (An Outline of Socioeconomic Policy in 1992), Mimeo, Warsaw, December 1991.

Chapter Six
Romania

Legacy and Change—Reform of Higher
Education and Restoration of Academic Work

Jan Sadlak*

Romania, like other former communist countries of Eastern and Central Europe, has to adjust its knowledge sector to its new political, economic, and social reality. The dimension and complexity of the required changes in such areas as internal organization, administration, and funding have been somewhat overshadowed by the well-justified "divine joy of freedom" following a victory over the particularly repressive communist regime and the refutation of the ideological straitjacket of Marxism-Leninism. But as the process of change continues, the crux of the "next stage" of postcommunist reforms is the setting up of conditions that will allow a reduction of the gap between Romania and highly industrialized countries in economic, technological, and scientific development. The academic community has given priority to a possibly fast restoration of academic freedom and the creation of an organizational basis for democratic self-governance. In this context, the roles of all parts of the knowledge sector, in particular higher education, can hardly be overemphasized.

The particular set of political, economic, and ideological circumstances imposed on Romanian higher education and science until the end of 1989 make it a unique case among the formerly communist European countries. There can be little doubt that the process of ongoing changes in postcommunist Romania has to take into account what has gone before, since those developments represent a legacy from which emancipation is, or should be, sought; they did allow the attainment of a certain level of educational and academic achievement and so cannot be entirely dismissed.

Particular attention is therefore paid in this paper to unraveling the consequences for higher education and academic work of the Ceausescu regime. Initiated in the 1960s as a departure from Soviet economic and foreign-policy domination and as an instance of the new receptiveness to Western technology and science, the moderniza-tion strategy was modified in the early 1970s and designated as the construction of a "multilaterally developed socialist society." Its con-ceptual framework for higher education and academic work was based on a unity of education, research, and production. This moderniza-tion program received general public support, particularly from the academic community, which saw it as a way of ending its isolation from the international academic community, as well as leading to an improvement of academic work conditions. It was also favorably per-ceived in the West, mostly for geopolitical reasons. After a few years of relative economic progress, however, the Ceausescu regime's modern-ization strategy rapidly turned into a self-destructive policy of self-sufficiency based on an incoherent mixture of rigid communist ideol-ogy; nationalism; strict adherence to central planning (which made it more an instrument of the state's control than an allocation system); and some randomly applied Western-style management solutions. Finally, it took on a particularly anti-intellectual "utopian" character that all but crippled Romanian academic life by reducing institutions of higher education to centers for training qualified laborers. As a consequence of this devastating legacy, Romanian higher education and research now require a systemic transformation. The changes in the organization, funding, and policy of higher education and sci-ence, which constitute this paper's second topical focus, were brought about since the fall of the Ceausescu regime.

Communist Takeover and Control of Higher Education and Science

The aftermath of World War II brought Romania into the geopo-litical setting of the Soviet-dominated Communist bloc. Even if Soviet military control together with economic difficulties left little room for the independent organization of life in postwar Romania, higher education and academic life functioned initially as in the precommunist years.

Romanian secondary and higher education, as well as the then few research institutions, were to a great extent a product of mid- and late-nineteenth-century educational reform, which coincided with the construction of the independent Romanian state. It was modeled on or inspired by the French system (and so the Napoleonic model of higher education), with an organizational structure based on universities, specialized institutions, and professional schools and with an educational philosophy which, at the polytechnic institute, combined classical humanism and advanced professional training. In a relatively short time, Romanian academics managed to establish high teaching and research standards in the exact sciences and medicine. The organization and certification of studies, names of degrees, and academic titles also give strong evidence of the French influence. Some titles, such as *baclaureat* and *licenta,* university degrees and *conferentiar* academic titles, continue in the Romanian educational system. This linguistic and cultural affinity led to preferential relations between Romanian students and academics and various Francophone academic and scientific institutions. Even though German culture, religion, and academic tradition were important in parts of Romania with a German ethnic minority (Transylvania), their influence was less pronounced among the Romanian universities than might have been expected.

Romanian higher education and science was remodeled by the educational reform of 1948. The "new vision" of Romania's educational system departed from the French-inspired model toward one based on Soviet educational concepts and practices. The new educational system was to be uniform and centralized, aimed at the needs of the socialist economy and new social order. Its ultimate objective was the formation of the "new socialist man," formed, in part, with a university-level trained "socialist intelligentsia." The graduates were perceived as the "intellectual cadres" of the new regime, readily contributing to its political, cultural, and economic objectives. Their formation required the introduction of new curricula together with close political supervision of students, academic teachers, and research workers, in order to ensure their adherence to communist ideology and the decisions of the communist party. The following principles characterized Romanian higher education and science until the end of 1989, even if some of the "borrowed solutions"—that is, some academic titles and the Russified study programs and textbooks—were abandoned in the early 1960s when Romania began to act more independently of the Soviet Union and to re-Romanianize the country's education and science content and organization:

1.   Higher education was both a functional instrument (for academic and professional training and development) and a place for political formation based on Marxism-Leninism and later supplemented by a variety of "theoretical" works attributed to Nicolae Ceausescu.

2.   Student enrollment and graduate employment were strictly correlated with centrally established manpower plans.

3.   Teaching and research activities of institutions of higher education were planned centrally, and academic interests were subject to the economic interests and ideological dictates of the communist party and the state bureaucracy.

4.   Disciplines were separated institutionally while their governance was controlled and coordinated with policy objectives by central political and state bodies.

5.   Research activities were narrowly application-oriented.

6.   Academic nominations and, to some extent, also higher academic degrees, required the final approval of the central accrediting body, which took into account political as well as academic criteria.

7.   A majority of the positions in the governing bodies of the academic institutions were held by the *nomenklatura,* thereby institutionalizing the party's controlling powers.

8.   Political control by "democratic centralism," by which academics, as party members, adhered to party discipline.

9.   Limited collective autonomy of academics and students.

10. Political coordination and supervision of the international relations of academic institutions and individual scholars.

These features characterized, with various degrees of adherence, practically all of the communist-controlled countries of Eastern and Central Europe.[1] But what made the case of Romania unique was the dogmatic implementation of this general model together with the personal power of Nicolae Ceausescu, and increasingly, from the mid 1970s, his wife, Elena Ceausescu, and their cronies.

Students

Among the East European communist countries, and despite political declarations of democratization of access to higher education, communist Romania was a country with one of the lowest participation rates of the student-age cohort in higher education. Higher education was not, at least after the late 1960s, regarded as an agent of social change. The number of students reflected political and policy decisions based on narrowly interpreted manpower planning, rather than changes in the student-age cohort or the state's response to high individual demand for higher education. In the 1960s, reinforcement of the meritocratic criteria by entry examinations was argued by the regime to be policy-determined and aimed, among other things, at equalizing entrance opportunities to university-level education.

The number of students was at its highest in the academic year 1979/80 when it reached 192,546 students (see Table 6.1), out of which 83.6 percent were full-time. In this category 14,438 foreign students represented 7.5 percent of the total; in the 1970s and 1980s, this was the highest among all communist countries. By the academic year 1987/88, the total number of students declined to 157,041 students, out of which only 57.6 percent were full-time (and this included 7,062 foreign students). The number of Romanian full-time students, a constituency essential for a viable academic community, returned to the same level in the late 1980s as in the mid-1960s.

The number of students in engineering and architecture was relatively high, particularly if compared with the number of students in other disciplines (see Table 6.2). The rate of increase was particularly high during the late 1970s, and only minimally declined in the early 1980s when the Romanian economy entered into a period of acute crisis and the "demand" for highly qualified manpower declined. As indicated by the figures in Table 6.2, this period of concentration on engineer training coincided with a marked decline, after 1975, in the number of students in the disciplines traditionally associated with the Romanian universities—law, humanities (including teacher training) and those disciplines in which academic work is carried out predominantly around basic and theoretical research, such as mathematics and natural sciences. The number of students in those disciplines was, in the late 1980s, similar to that in the early 1950s. The long-term consequences of this decline in training and research capacity in those disciplines in which Romanian academics were internationally recognized were clearly disregarded.

In the mid-1970s, students in engineering and architecture represented between 30 and 40 percent of all students. The technological orientation of the student enrollment continued, so that in the

1980s more than 60 percent of the student population was registered in engineering (64.6 percent of the total student population in
the academic year 1989/90), while the extent of humanities and
natural-sciences education was substantially reduced. The number
of students dropped significantly after 1975 and in the fine arts became dismally small after 1980—less than 1 percent (see Table 6.2).
No other communist European country at that time so narrowly
interpreted the role of its system of higher education.

Like other countries, Romania experimented with various organizational and institutional forms of accelerated-degree and shorter-
than-university-level study programs. They were introduced under the
banner of democratization of access and as a new way of responding
to demand for certain graduate skills, usually by engineers and teachers. Moreover, vocational orientation was emphasized to the detriment of the academic process in acquiring knowledge and professional
qualifications. These programs, unlike full university-level studies, did
not require 4.5 to 6 years of full-time course work, rigorous periodic
and final examinations, and diploma work. The schemes introduced
in Romania in the academic year 1970/71 mainly provided technical
studies—three years for full-time and four years for part-time students—in order to train sub-engineers (*subinginer*) and subarchitects
(*conductor-architect*) who were expected to function at an intermedi-
ate level between that of fully trained personnel and the foreman on
the factory floor or construction site. They were awarded a diploma of
"subengineer," based on vocational training rather than theoretical
study. Instruction in the latter courses of study was provided mainly at
the institutes for subengineers (*Institutul de subingineri*), which had a
lower academic status than the five-year engineering courses provided
by polytechnics, technological institutes, and some universities. In the
1970s, the students in the shorter courses accounted for more than 40
percent of all students in engineering. These figures sank to less than
15 percent in the 1980s, thus showing that this category did not greatly
attract either students or their potential employers.

The relatively large number of students in medicine was a result
of the number of foreign students. Paid in hard currency, tuition
fees represented an important incentive for Romanian policy-makers. Their numbers declined, however, by some 50 percent after the
academic year 1981/82, largely because of poor living conditions.

Table 6.1 Number of Students (by Form of Study and Foreign Students) and Academic Staff in Romania in the Academic Years 1948/49 to 1992/93

Academic Year	Total No. Students	Full-Time Students	Part-Time Students	Foreign Students	Academic Staff
1948/49	48,676	48,676	—	—	5,638
1949/50(a)	48,615	45,816	2,799	—	7,088
1950/51	53,007	46,195	6,812	—	8,518
1951/52	61,123	49,010	12,113	357	8,917
1952/53	71,513	52,434	19,079	526	8,469
1953/54	80,593	59,935	20,658	750	7,866
1954/55	78,860	61,948	16,912	833	8,278
1955/56	77,633	60,347	17,286	877	8,369
1956/57	81,206	56,170	25,036	986	8,154
1957/58(b)	80,919	51,094	29,825	920	8,982
1958/59	67,849	45,501	22,348	862	8,009
1959/60	61,980	44,774	17,206	953	8,041
1960/61(c)	71,989	56,409	15,580	897	8,917
1961/62	83,749	68,338	15,411	741	10,360
1962/63	98,929	79,626	19,303	869	10,753
1963/64	112,611	88,478	24,133	617	11,965
1964/65	123,284	93,992	29,292	503	12,465
1965/66	130,614	96,407	34,207	439	13,038
1966/67	136,948	98,315	38,633	729	13,404
1967/68	141,589	99,688	41,901	971	13,792
1968/69	147,637	102,507	45,130	1,261	12,950
1969/70	151,705	106,302	45,403	1,562	13,166
1970/71(d)	151,885	107,437	44,448	1,766	13,425
1971/72	148,428	106,832	41,594	2,074	14,470
1972/73	143,985	103,171	40,814	2,041	14,488
1973/74	143,656	103,779	39,877	2,287	14,537
1974/75	152,728	108,750	43,978	3,873	13,931
1975/76	164,567	115,769	48,798	4,971	14,066
1976/77	174,888	127,225	47,663	6,677	13,662
1977/78	182,337	138,500	43,837	9,367	13,575
1978/79	190,560	148,859	41,701	11,782	14,227
1979/80	192,546	—	—	14,438	14,503
1980/81	190,769	161,110	31,659	15,888	14,592
1981/82	190,903	157,708	33,195	16,962	14,354
1982/83	181,081	142,418	38,663	16,251	13,931
1983/84	174,042	127,878	46,164	14,808	13,344
1984/85	166,328	112,803	53,525	13,068	13,250
1985/86	159,798	100,040	59,758	10,774	12,961
1986/87	157,174	93,287	63,887	8,897	12,504
1987/88	157,041	90,490	66,551	7,062	12,036
1988/89	159,465	92,091	67,374	6,503	11,810
1989/90	164,507	94,952	69,555	6,669	11,696
1990/91	192,813	136,035	56,778	—	—
1991/92	225,794	159,678	55,548	10,568	16,129
1992/93	237,521	177,584	48,937	11,000(e)	20,810

Source: Data collected from the Romanian statistical yearbooks and other publications.
(a) Part-time studies were introduced.
(b) Academic year 1957/58 includes also teaching and laboratory assistants.
(c) Three-year teacher-training studies classified as equal to other university degree studies.
(d) First year of student enrollment to three-year (four-year for part-time students) form of technical studies.
(e) Approximate number.

There were only a small number of students in agriculture, a surprising fact given the importance of agriculture to the Romanian economy and Ceausescu's declaration of the "agricultural revolution." Their numbers in the academic year 1987/88 declined by 37.2 percent compared with the academic year 1980/81. The substantial differences in the level of social and cultural services provided in urban and rural settings explain why many graduates of agriculture and other fields sought employment, after a period of compulsory job assignment, in big cities, and were even ready to accept jobs not related to their field of study. In response, administrative measures, introduced in 1981, were intended to ensure that agricultural graduates would work on the land or in jobs directly related to agricultural production or enterprises. The actual result of these pressures was to reduce the number of applicants for studies in this field.

The evening and correspondence courses that were introduced in Romania in the academic year 1949/50 served mainly as a "solution" to the need to train a larger number of people rapidly. For this reason a "would-be-student" could apply for admission in a field similar to the job already held. Correspondence studies became the dominant form of part-time studies from the late 1950s until the early 1970s, while, in the 1980s, evening courses were preferred for organizational and academic reasons, representing about 70 percent of all part-time students (see Table 6.1).

Regarding the last years of communist rule in Romania, three further observations illustrate why the educational orientation toward the "world of work" could be considered excessive, particularly in comparison with other European communist countries.

First, the number of study places and the compulsory two- to three-year postgraduation job assignment was rigidly linked with national manpower plans. It officially rested on a contract signed by the students at the beginning of their studies. This arrangement apparently avoided regional shortages of specialists. But it can also be argued that it was another mechanism used by the regime to exert control over the intelligentsia, as virtually all professional posts were assigned by the state.

Second, the principle of "relating higher education to economy," applied with zeal, resulted in a highly specialized curriculum. The regime justified this focus for university-level education by saying

Table 6.2 Students per Major Group of Disciplines (Selectively in the Academic Years 1950/51 to 1989/90) in Romanian Higher Education

A = Total number of students in the given discipline
B = Percentage of students in the given discipline (compared with the total number of students)

Discipline	Engineering		Agriculture and Veterinary		Economics and Management		Law and Public Administration		Medicine		Humanities and Teacher Training		Mathematics and Natural Sciences		Fine Arts		Others		Total
	A	B	A	B	A	B	A	B	A	B	A	B	A	B	A	B	A	B	
1950/51	13,344	25.0	7,677	14.3	11,650	8.8	2,964	4.6	7,610	14.4	15,056	28.6	2,129	4.0	269	0.3	—	—	53,007
1955/56	30,375	39.1	7,661	9.3	5,921	7.6	3,327	4.3	9,796	12.6	19,042	24.5	1,160	1.5	351	0.5	—	—	77,633
1960/61	22,190	30.8	8,306	11.5	5,085	7.1	3,101	4.3	7,825	10.9	14,615	20.3	9,137	12.7	1,730	2.4	—	—	71,989
1965/66	42,304	32.4	9,961	7.6	12,866	9.8	4,534	3.5	9,345	7.2	38,208	29.3	10,977	8.4	2,419	1.8	—	—	130,614
1970/71	44,518	29.3	9,074	6.0	21,016	13.8	5,901	3.9	9,898	6.5	43,207	28.5	14,901	9.8	3,370	2.2	—	—	151,885
1975/76	64,088	38.9	11,962	7.3	22,854	13.9	6,820	4.1	17,008	10.3	21,568	13.2	17,487	10.6	2,780	1.7	—	—	164,567
1980/81	113,323	58.8	10,683	5.6	21,919	11.4	3,863	2.0	23,381	12.1	8,720	4.5	8,673	4.5	2,207	1.1	—	—	192,769
1985/86	99,779	62.4	7,005	4.4	16,485	10.3	2,380	1.5	18,833	11.8	6,261	3.9	8,156	5.1	899	0.6	—	—	159,798
1989/90	106,299	64.6	6,886	4.2	15,493	9.4	2,362	1.4	16,763	10.2			15,828	9.6	936	0.6	—	—	164,507

Source: National statistical yearbooks and author's calculations.
*Figures are consolidated for the Humanities and Teacher Training and Mathematics and Natural Sciences disciplines.

that "the university [in Romania] must serve a society for its advancement by educating useful specialists. We can no longer give a broad culture without any purpose."[2] It was also this "pragmatic" vision of higher education that brought about the decision to reduce student enrollment not only in fine arts but also in the humanities and natural sciences. This exaggerated specialization also affected secondary education in which vocational schools and specialized lyceums dominated to the detriment of general, academically oriented curricula.

Third, since the late 1970s, study programs were organized not only according to the needs of professional training, but also to develop students' "interests in and love for work" as well as "to associate theoretical knowledge with the requirements of productive and research activities."[3] Other communist countries of the region may have had the goals of polytechnical education, work experience, and technology transfer mechanisms, but the uniqueness of the Romanian approach lay in "for work through work"; both students and staff were required to undertake "productive activities." Each school was supposed to be at the same time a "place of learning" and a "unit of production." For this purpose, three categories of production units were established: "enterprises," which had a legal identity and which functioned on the same principles as any other business firm; "workshops or design units," which functioned without independent legal status; and "production sections and laboratories," with a limited production or design activity. All of them, independent of their formal status or size, had their production plans and contracts, which were mainly the responsibility of the academic staff. In addition, it was required that subjects for student papers, diploma papers, doctoral theses, and academic research projects would deal with concrete problems in factories, farms, and other sectors of economic or social activity.[4]

Academic Staff

The quality of academic work in relation both to teaching and research greatly depends on the intellectual and moral quality of the academic staff. The communist regime in Romania was well aware that it had to keep a close watch not only on the content of study programs but also on the selection process for academic personnel, particularly the two "senior" rank positions of professor (*profesor universitar*) and associate professor (*conferentiar*), even if by law they could be occupied only through open competition and required the possession of an advanced academic degree of doctorate.[5]

Three bodies of the higher educational establishments occupied themselves with academic appointments: a search committee for the particular appointment (*Comisia de concurs*), the academic board of the faculty, and the academic senate. The first assessment of formal qualifications was carried out by the search committee. Its conclusions and recommendations were then submitted to the academic board of the faculty, which transmitted them, together with its own recommendations, to the senate. The decision of the senate was in turn submitted to the rector, who had to present it to the Ministry of Education. Officially, the ministry was responsible only for ensuring that proper procedures had been followed. In fact, the ministry examined the substantive assessments and recommendations jointly with the personnel department of the central committee of the Romanian Communist Party. From 1975, this body was under the direct supervision of Elena Ceausescu. It was only after this review that the Ministry of Education presented its decision to the institution.

Appointments to "lower" academic positions (lecturer [*lector*], project supervisor [*sef lucrari*], assistant and junior assistant [*preparator*]) were subject to less political scrutiny, and nomination could be decided by the faculty and administration of the academic institution. An important instrument for ensuring obedience in those aspiring to an academic career was created by the law of 1948, the High Commission on Diplomas (*Comisia Superioara de Diplome*). Its creation, copied from the Soviet academic model, was officially justified by the need for the preservation of the quality of doctoral degrees. But its most important function, operating under the immediate jurisdiction of the Minister of Education, who was also its president (the deputy minister in charge of higher education was its vice president and the scientific secretary), was its role as a kind of high court guaranteeing political control over the granting of doctoral degrees awarded at the institutional level and so confining such decisions within the standards and desires of the Communist Party. When, in the late 1970s, the communist regime dropped any remaining elements of its more liberal approach to higher education embodied in the law of 1968, even the undertaking of doctoral studies became subject to prior approval by the top-level political bodies. The new law of 1978 represented the legislative confirmation of closer political control over academic work.[6] It put the State Council, in reality its chairman, Nicolae Ceausescu, in a prominent role in two important matters of academic policy. It was the council that established the yearly student matters of academic policy, and it was the council that established the yearly student enrollment figures, following the proposals of the Ministry of Education that defined the conditions under which each doctoral degree could be awarded. Until then, it had been the universities and other doctoral-degree-awarding institutions that—within the total limits established by the ministry—had made such decisions. The autocratic and whimsical views about the need for training academic staff resulted in a situation where, by 1989, the total number of doctoral students had decreased to the dismal level of 652 persons.[7]

The academic staff were subjected to a bureaucratic work discipline. On average, they were required to spend five eight-hour work days per week at their respective institutions or at other places closely linked to their academic activities. The distribution of those forty hours were as follows: eighteen hours teaching, tutoring, giving examinations, and so forth; twelve hours on research and related activities; and four to six hours for other activities having to do with students.[8] These regulations were frequently infringed upon because it was almost impossible for the bureaucracy either to keep teachers under strict surveillance or to make a clear distinction between the various tasks. Academic staff also were occasionally expected to participate in or supervise "voluntary patriotic labor" projects, which were mostly harvest-season drives and political gatherings.

The legislation regulating the status of the permanent teaching personnel in all types of educational institutions in Romania, adopted in March of 1969, did not make any specific reference to tenure. All categories of teaching personnel in academic institutions—professor, *conferentiar*, lecturer, and assistant—were employed on the basis of an unlimited contract. However, junior academic staff had to show, in the course of the first six years, that they had achieved sufficient progress in their teaching and research activities. Their contract could

be revoked by the Ministry of Education on the basis of a "motivated proposal," made by the chairman of the department, confirmed
by the academic board of the faculty, and approved by the rector of
the given institution of higher education.[9] The contracts of senior
academic staff could be revoked according to regulations stipulated
in the labor law, but their positions were fairly stable unless they had
been singled out for political reasons. In such cases they were banned
from teaching and relegated to obscure tasks within the same institution, as were those academics who openly tried to seek emigration. In some extreme cases, they were forced to take employment
outside academic centers.

The number of academic staff in all Romanian institutions of
higher education fell from more than fourteen thousand in 1975 to
fewer than twelve thousand in 1989/90, which more or less corresponds to the number in the early 1960s, when student enrollment
was substantially lower (see Table 6.1). The student-teacher staff ratio suffered accordingly. In fact, very few academic appointments
were made under the Ceausescu regime, especially if compared with
student enrollment trends. This was the result of the government's
financial straits and the indifference, if not hostility, of the ruling
group toward the academic community, perhaps with the exception
of a relatively small but influential group of cronies. A sad irony of
this situation was that despite Elena Ceausescu's pretense of seeing
herself as a scholar of international importance, reinforced by duti-
ful references to the great "theoretical and practical contribution" of
Nicolae and Elena Ceausescu to "development" of education and
science in Romania, the general material situation of higher education was bad. It would be true to say that, especially after 1980, the
conditions for academic work were humiliating and respect for academic freedom was virtually eliminated.

Science Policy and Conditions for Academic Work under Ceausescu

Lenin once said: "We must take control over science, technology,
knowledge and art. Without these we shall be unable to build a
communist society."[10] Similarly, the development of education and
scientific research was declared by Nicolae Ceausescu on numerous
occasions to be one of the major factors in the modernization of the

national economy and so made an objective of party and state policy. In 1965, Ceausescu warned that lagging behind in science "can only have negative consequences . . . and can lead to dependence on foreign countries."[11] From then on, practically all major documents of social and economic development stressed the role of the knowledge sector. For example, the earlier plans emphasized various branches of heavy industry, while the 1976–1980 National Unified Plan on the Country's Socio-Economic Development regarded progress as a result of the "scientific and technological revolution" that would encompass higher education to ensure qualitative and quantitative training in order to realize the extensive application of research findings in all socioeconomic activities.[12]

According to the official report that evaluated the implementation of this plan, the adopted science policy was more than successful. It reduced the overall duration of the research-project production cycle and produced more than 10,300 new types of machines, products, and installations, more than 9,000 new or modernized technologies, and more than 5,100 new or modernized consumer products. Moreover, as a result of this applied research, the evaluation report claimed there had been substantial progress in basic and prospective research in a wide range of disciplines.[13] However, there is some question about the reliability of these claims.

Indirectly, academic institutions were also part of the Romanian strategy to achieve technological progress by access to Western technologies and manufacturing expertise in such areas as nuclear energy, based on the Canadian Candu nuclear power plant, or aviation, through a joint venture to build the British-designed civil midsize plane the BAC-111. There were also some ambitious projects to develop military technology, like that of building a fighter plane in cooperation with Yugoslavia. This technologically driven modernization strategy temporarily improved the possibility for Romanian academics, even if under the close scrutiny of the state authorities, to participate in international academic activities and such initiatives as the Fulbright program, especially in those disciplines relevant to advanced technologies.

The above examples well illustrate that the communist regime in Romania had quite a clear picture of the strategic need for scientific development. But it tended to implement this under strict institutional control and to give its political and material support exclusively to the type of research that corresponded to its ideological and narrowly interpreted economic objectives. The Romanian strategy of scientific research for the 1980s foresaw that twenty-five hundred out of some three thousand research and engineering projects would relate to the manufacturing and construction industry.[14] The level of research funding in other industries, for example telecommunications, was very limited by comparison. Telecommunications R&D funding represented only some 0.5 percent of the funds allocated to R&D in the manufacturing and construction industry. However, the policy regarding communications is consistent with a regime that required the registration of typewriters at the local police station.

A different approach was made in early 1970 to the coordination of this "science policy." Two state bodies—the National Council for Science and Technology (*Consiliul National pentru Stiinta si Tehnologie*) and the Academy of Social and Political Sciences (*Academia de Stiinte Sociale si Politice*)—were directly subordinated to the central committee of the party. Both organizations played an important role with regard to the research activities carried out in higher educational institutions. The first of these organizations was the principal body responsible for the development and implementation of Romania's technology research strategy. It was also administratively responsible for the Academy of the Romanian Socialist Republic (*Academia Republicii Socialiste Romîne*) which, after some reorganization in 1970, had replaced the Soviet-styled Academy of the Romanian People's Republic (*Academia Republicii Populare Romîne*). The fifty research institutes and laboratories were dissolved or transferred to industry or to institutions of higher education.

After this restructuring, the Academy of Social and Political Sciences (*Academia de Stiinte Sociale si Politice*) became responsible, among other things, for the coordination of research activities in the social sciences. This change was unfavorable to higher education in general, particularly universities. Universities had to coordinate their social-science research activities with the academy. The academy's teaching "branch" was carried out by the party school—the Academy for Training and Development of the Leadership Cadres (*Academia' Stefan Gheorghiu' pentru Pregatirea si Perfectionarea*

Cadrelor de Conducere)—which not only became an academic institution and so an integral part of the system of higher education but
also an elitist training center for the party, trade unions and state
bureaucrats, diplomats, journalists, and teachers of social and political sciences. Under such an organizational arrangement and "academic leadership," the social sciences in Romania had reached an
unimaginable level of mendacity. The serial production of apolo-
getic "academic works" was, of course, detrimental to the academic
credibility of the social sciences. The moral consequences of this
intellectual servility are still visible today.

Elena Ceausescu: A Supreme Autocrat over Romanian Science and Education

An understanding of the powers held by Elena Ceausescu is useful
for an examination of the conditions for academic work in Romania
since the mid-1970s. The first time that she appears to be directly
involved in academic matters was when she became chairperson of
the above-mentioned National Council for Science and Technology. In the years following her nomination, she quickly became a
supreme autocrat over Romania's academic and science policy. She
demanded to be addressed as "Academician, Doctor Engineer" always. Romanian diplomats and academics obligingly responded to
her desire to be awarded with honorary degrees, titles, and to have
Romanian and foreign publishers, academic institutions, and societies concede to her wishes. She and her supporting watchdogs
showed an almost complete disregard for academic achievement and
research ethics; this permitted nonmeritorious promotions which
damaged the image of academia and the credibility of academic work.

The source of her influence over the Romanian academic community lasted for many years. She was a member of the top party
and governmental bodies where she was in charge of personnel policy,
including the political approval of nominations for academic and
administrative appointments. She became involved, too, in such
matters as the foreign travel of individual academics. Her "pet project"
was the NIC, or the National Institute of Chemistry (*Institutul National de Chimie*), formally attached to the Bucharest Polytechnic
Institute. It was set up in the early 1970s on the basis of her "innovative conception . . . which provided for the elaboration of origi-

nal, high-tech methods of competitive technical-economic parameters, as well as for the manufacture of basic chemical products to serve the development of the other branches of the national economy" [*sic*].[15] It was a unique academic institution. In the late 1980s it included two faculties (chemistry and chemical engineering), twenty-seven specialized research centers and production units of the Ministry of Chemical Industry, as well as the few remaining small chemical research units that were under the Academy of the RSR. The institute's scope and prerogatives directly affected academic work and other institutions, as it coordinated the chemistry departments of all institutions of higher education. The creation of the NIC directly affected the structure and the academic profile of Romanian universities by dissolving their faculties of chemistry and transferring their staff to institutions with applied research agendas. The University of Bucharest, for instance, saw its faculty of chemistry eliminated and its staff and equipment moved to the institute, while at the University of Jassy, the chemistry faculty, one of the most prestigious in the university, transferred in 1974 to a new faculty of chemical engineering and technology at the Polytechnic Institute in Jassy. No other central research institutes—the Central Institute of Biology or the Central Institute of Physics—was equivalent to "her" institute in its academic prerogatives, available funding, or number of research personnel.

The apogee of her formal control over Romanian academic life was the chairmanship of the National Council of Education and Science (*Consiliul National al Educatiei si Stiintei*), created in 1985 to reinforce the policy of directing research toward industrial applications. As a result, the total number of faculty members in establishments of higher education dropped from 132 at the beginning of the 1980s to 101 in the academic year 1985/86, with mathematics, agriculture, arts, and medical faculties hardest hit. By then, most of the basic research in natural sciences had been eliminated from institutions of higher education, and research in the humanities had to be given, sometimes quite artificially, a practical relevance in order to comply with the narrow doctrine of "productive" academic research. Only 10 percent of the total research budget was permitted to be allocated for "fundamental" research. The policy assumed that institutions of higher education, by charging fees for undertaking research, design, and production activities, would acquire funds for their own research and scientific equipment. Apparently through such "self-financing" measures, the Bucharest Polytechnic Institute was able to raise "over one-third of the annual budget needed for the development and the equipment of laboratories."[16]

The damage done to the academic integrity of the Academy of the RSR by a combination of policy and the autocratic personal decisions made by E. Ceausescu was severe. Its main task was to fulfill ceremonial and honorific academic functions as well as to host and take part in international scientific conferences that received the approval of the regime. Under the Ceausescu regime, academic research was not "placed in the 'golden cage' of the Academy of Sciences,"[17] nor was there any meaningful competition for funding or academic prestige between institutions of higher education and research organizations of the Academy of Sciences.

Learned and Scientific Societies

An important indicator of the vitality of the scientific community and conditions of academic work is the situation of the learned and scientific societies. At the end of the 1980s there were some forty learned and scientific societies that, officially at least, were not directly dependent on the state. Their number and membership was small, even in comparison with other socialist countries where such organizations often permitted the intellectual and material survival of nonconformist academics. In Romania, if academics exercised a similar choice, they would be banned from any possibility of academic work. The Romanian learned societies were very small, with fewer than one hundred members. Many, particularly those in medical and technical sciences, represented a long tradition going back to the end of the last century, and they played an important role in the foundation and early development of higher education in Romania. However, their role, particularly during the Ceausescu era, was modest and limited to minor aspects of scientific and professional communications, such as, for example, the sponsorship of small, irregularly published journals.

Particularly adverse conditions for education and science marked the late years of the Ceausescu regime. Despotic rule resulted in egregious human rights abuses that directly affected some members of the academic community. The "revolutions" and megalomaniacal construction projects, coupled with accelerated repayment in the late 1980s of the country's foreign debt, brought general privation. The policing of academia eliminated the critical role of free academic inquiry from public debate. Unfortunately, frequent tacit complacency on the part of the members of the academic community neither reduced the restriction on academic work nor improved the material situation of higher education in general.

That is not to say that there have not been any concrete scientific achievements or good study programs in Romanian higher education. Despite all the restrictions, control, and dismal physical resources, there was some genuine academic work done. This kind of encouraging "surprise" was confirmed in a scientometric survey carried out in 1988 by a team from the Philadelphia Institute for Scientific Information. While they documented an improvement in Romania's position in basic and life sciences based on papers in those disciplines internationally cited in the later 1970s, they were unable to explain why this was the case, given the policy of the Ceausescu regime to systematically destroy free enquiry and independent academic work.[18] There were also rare cases in which the academic community was successful in obstructing Elena Ceausescu's plans to "improve" higher education, as, for example when she intended, in the mid 1980s, to collapse all types of artistic and fine arts institutes into "all-arts academies."

These exceptions do not change the overall conclusion that the instrumental use of academic work, particularly when combined with the grotesque science politics of Elena Ceausescu, completely failed. In the early period of Ceausescu's modernization policy, more cooperative relations toward the academic and science community might have been attempted. Gradually, the politically and personally motivated restrictions were increased, coupled with a lack of recognition for even a limited degree of professional autonomy in academic work. Higher education was treated as an instrument for the policies and tasks set by the Ceausescus and their cronies. The distortions in social sciences and arts were particularly grotesque.

The Road Ahead—Challenges and Responses

Romania enters its postcommunist era as a country with one of the most impoverished populations in Eastern and Central Europe. Higher education and other parts of the knowledge sector will have to position themselves as a contributing party willing to establish a market economy and to reinforce a civic and democratic society. At the same time, higher education will have to face competition for scarce public resources, which are badly needed for restructuring and development.

Political and Social Conditions for Educational Reform and Academic Renewal

The role played by students and some academics as the vanguards and intermediaries of the leaderless revolution in December of 1989 (which brought about the fall of the Ceausescu regime and the abolition of communist rule in Romania) should be considered a particularly positive factor in forming a new perception of the place of the academic community in Romanian society. The positive role and sacrifices of the student population and some members of the academic community have helped to bring back some respect for moral standards in their collective behavior.

To some degree, however, the academic community, especially the students, has become a "victim" of its success. The most disturbing incident was the miners' brutal attack on students and the savage destruction of academic buildings and equipment in Bucharest in June of 1990; it met with strong criticism from the intellectual and educational community and some opposition politicians.

Another contentious issue is ethnic tensions, principally between Hungarians and Romanians, which have direct repercussions for the functioning of higher education and academic life in general. For the time being, it appears the present day political leadership does not want to take a political risk by recognizing the demands of the Hungarian minority to found its own university. The reinstating of a Hungarian-language university,[19] originally scheduled for the autumn of 1990, has been postponed, because it would, according to the government, revive past dissensions between Romanians and Hungarians in Transylvania. By taking this short-term position, the present political leadership confirms the view that its strategy involves a limited form of political pluralism; and this also applies to higher education for minorities. Nor does it show a political understanding that such a decision can have adverse consequences and result in the further deterioration of relations between various ethnic groups in the whole Balkan region.

It should, however, be recognized that since 1992 there has been some progress in this sphere. The enrollment of students of Hungarian origin has increased. It now represents about 4.5 percent of the total student enrollment. In addition, there are no restrictions on study abroad, which, among other things, makes it possible for Hungarian students to study privately in Hungary.

The most important political change in Romanian academic life is "deideologization"; this has led to the restoration of an academic framework and self-governance. Ideologically motivated names added to the official names of the institutions of higher education were dropped. More important were the content changes of almost all study programs, with the deepest changes introduced in social sciences and economics. The party school, the Academy for Training and Development of the Leadership Cadres, in Bucharest was closed. At the same time, new faculties of economics have been created within the universities in Galati, Brasov, and Constanta. Finally, theology, which was eliminated from the universities in 1948, has returned as a university discipline.

The deideologization of education and science in Romania also has direct repercussions for its international academic relations. Romanian academics have shown considerable initiative in reestablishing their links with the international academic community. An international conference on academic freedom and university autonomy, held in Romania in May of 1992, marked "a ceremonial return to the European academic community." But this was also a forum in which it was made clear, once again, that the process of change is bringing much hardship to the academic community; as was underlined by the president of the newly created National Rectors' Conference of Romania, "East European universities need overseas contacts to rejoin the European space."[20]

As an academic community, Romania, like the other newly capitalist countries of East and Central Europe, has to be aware that self-governance and institutional autonomy have a price; both students and academic staff now must see themselves in terms of the often hard rules of the market economy. One major problem is the lack of general social acceptance of the cost-recovery principle, such as tuition fees for state-funded institutions, combined with the difficulty of finding alternative sources of funding for higher education. The present structure and ownership of the Romanian economy are neither sufficiently diversified nor "prosperous" enough to present any meaningful alternative. Foreign investments are lower than in many other countries of Eastern and Central Europe,[21] and those foreign companies that decide to invest in Romania rarely supply sufficiently advanced technology—with the possible exception of the pharmaceutical industry, which might provide a number of opportunities for advanced training or research. Consequently, the state budget for higher education and scientific activities remains the almost exclusive source of funding, even though the Romanian government faces considerable financial difficulties. In this situation even the few existing locally based foundations, whose scope is to support education and scientific activities, represent a welcome beginning.

New Initiatives, but also New and Old Problems

The difficulties of the postcommunist countries in forming mature market relations, combined with everyday hardships, widespread disillusionment among students and intellectuals, and frequently ineffective politicians, mean that current political and economic conditions are not particularly conducive to the process of reform in higher education. The present situation in Romania confirms this general observation. Nevertheless, the new economic reality is gradually changing relations between higher education and the newly restructured Ministry of Education, which has replaced the previous arrangements by which higher education and science were administered by two separate governmental bodies. The new arrangement is showing greater transparency and respect for institutional autonomy at a time when Romanian higher education is experiencing a period of structural transformation, part of which has been quantitative expansion of the student population, staff and institutions.

The student enrollment in public institutions of higher education increased to 192,813 (136,035 full-time and 56,778 part-time students) in the academic year 1990/91, and it grew in the next two years to more than 237,500 students in the academic year 1992/93 (see Table 6.1). The number of full-time students grew substantially, while the proportion of part-time students dropped. An expedient political decision to satisfy high social demand for full-time university-level education, which was suppressed by the previous regime, has played a certain role in this development.

In addition, there are some 80,000 students in private higher education institutions. Even if this figure could be considered high and there had been a noticeable leveling off of the social demand for higher education in general, there is no doubt that the size and composition of the student population in Romania has dramatically changed.

As in previous years, most programs require four to six years of study, depending on the subject area. They can be followed by postgraduate studies leading to the doctorate, which requires an additional three to four years of course work and the preparation and public defense of a thesis. Finally, many institutions of higher education provide, in their "university colleges," short-cycle, three-year study programs. Thus, the structure of study programs in Romanian higher education, *groso modo*, is evolving toward a system in which bachelor's, master's, and doctoral degrees are awarded.

A radical change took place in the number and institutional structure of higher education in Romania as a consequence of the startling increase in the number of higher education institutions adopting the status of "university" and the massive emergence of private establishments.

Before analyzing those developments it should be pointed out that, in Romania, particularly in the mid-1960s and early 1970s, universities were required to accommodate technological, agricultural, and medical faculties. This was different from the majority of communist countries where these disciplines were assigned to independent, specialized institutions.

This historical background, together with local ambitions and a legislative vacuum in the early days after the collapse of the communist regime, explains the sudden increase in the number of higher education establishments calling themselves "universities." (In the course of the years from 1991 to 1993 the number of state universities increased from 8 to 36.)

At present, out of a total of forty-eight state institutions of higher education, there are: thirty-six various types of university, one institute of technology and construction, one medical academy, one school of architecture, one national school of physical education and sport, one national school of political and administrative studies (which offers only postgraduate study programs), one institute of civil navigation, two theater and film institutes, two music academies, one academy of fine arts, and one academy of fine arts that also has a faculty of music. There are also five higher educational establishments that are directly administered by the Ministry of National Defense and the Ministry of Internal Affairs, as well as some schools for priesthood functioning under the general administration of the State Secretariat for Religious Affairs.

The emergence of private higher education has been mainly guided by the interests of those who perceived a need in postcommunist Romania for an alternative to the state-controlled institutions and a response to individual and social demand for higher education. Proponents of private higher education also point out that less rigid criteria for formal academic qualifications allow teaching to be undertaken by experienced practitioners, who for political or professional reasons, have not been able to follow a traditional academic career pattern. Not less important is the fact that teaching in the private institutions can represent a considerable additional income to notably impoverished (in recent years) teachers in the state higher education institutions.

There is no doubt that private higher education is nowadays an important component of the "learning market" in Romania despite valid concerns regarding the academic level and professional quality of programs provided in such institutions. A substantial number of them, however, should be considered as undergraduate training colleges or postsecondary vocational schools, rather than universities.

The need for a resolution of this issue (as well as reservations concerning the academic viability of some newly created public higher education establishments) led, after several months of heated debate, to the adoption by the Romanian parliament on November 16, 1993, of the "law on the accreditation of higher education institutions and recognition of diplomas."[22]

The law lays down procedures for the functioning of state and private higher education institutions. All institutions, regardless of the date of foundation and "property" status, must function as nonprofit organizations. They must also undergo a periodic academic evaluation that is a prerequisite for obtaining or preserving the status of an "accredited" institution of higher education. With regard to university-level institutions, the law defines this status as the right to organize exams for *licenta* (a diploma obtained after four to six years of university-level studies) and to issue diplomas recognized by the Ministry of Education. Interestingly, the right to organize doctoral studies has not been guaranteed to all accredited institutions but only those which "meet the conditions established by the law." The law has only confirmed that there are significant differences with regard to the academic and research potential among university-level institutions in Romania.

In view of the massive use and abuse of the title "university," the Romanian legislators opted for legal regulations to cover use of this title. It can be used only by those institutions that have been accredited or have received an "authorization." The law has automatically granted accreditation status to those institutions that were founded before December 22, 1989, the date of the events that culminated in the fall of the communist regime. Thus, indirectly, the law has attested the good academic standing of all university-level institutions that were established before the communist regime was overthrown, even if later on they, too, will have to undergo a process of academic evaluation. This process is compulsory every five years, for each specialization, faculty, and university college of a given higher education institutution.

It is now much harder for the new institutions to obtain accreditation. For them, the law lays down two stages: an "authorisation for temporary functioning" allowing only teaching activities, and second "accreditation." All institutions—state and private—that were founded after December 22, 1989, have been obliged to seek temporary authorization within six months of the date of publication of the law. The law also sets quantitative, quite demanding, minimal standards for accreditation in such areas as teaching staff, study programs, facilities, research, and financing practices.

The main body that is responsible for implementing the law is the National Council for Academic Evaluation and Accreditation (*Consiliul National de Evaluare Academica si Acreditare*). It is composed of 19 to 21 members nominated by the Parliament, on the proposal of the government, for a period of four years. One third of its members will be replaced each term. The council has a right to establish evaluation commissions, members of which are selected from candidates proposed by the professorial councils that function at the faculty level, as well as other specialists. The council has become an important factor in determining many aspects of Romanian higher education.

The State of Mind of the Romanian Academic Community
In the academic year 1991/92, the number of academic staff increased to 16,129, and in 1992/93 to 20,810 (see Table 6.1). This increase was due to higher-education institutions' regaining control over the selection and promotion of academic staff. There has also been a renewal of the academic staff by opening some forty-three hundred positions for appointment and promotion, mainly the two "senior" rank positions of professor and *conferentiar*. Such a massive increase in hiring shows that the academic community has managed to convince the government to provide, despite financial constraints, the required budgetary resources. This laudable development, if carried out with due respect for required academic qualifications, touches another issue that has already received some international attention— that of the academic "brain-drain" from the countries of East and Central Europe. This problem does not represent, at least for the time being, a cause for concern in Romania the way it does in some of the other countries of the region.[23] Nevertheless, if the conditions for academic work do not improve, for example by greater access to scientific information (computer-based networks, publications, research facilities, and travel, as well as the overall conditions of life), the problem of an academic brain-drain, particularly in technology-related fields, could become an important issue.

Interestingly, there was a comprehensive and quite representative study of the "state of mind" of the Romanian academic community, students, and academics, conducted at the end of 1991 in five principal academic centers in Romania—Bucharest, Jassy, Cluj Napoca, Timisoara, and Galati. The general picture that emerges from the survey shows that despite all the problems in the political life of the country, the restoration of academic freedom has made a lasting imprint on the contemporary system of higher education and the organization of academic work in Romania. Thus, the present agenda of reforms and the forthcoming new law on higher education should, among other things, deal with the issues of functional university autonomy and supporting institutional and organizational arrangements in the areas of elaboration and validation of study programs, academic appointments and titles, budget and its administration, collective representation of the academic community, and division of policy and decision-making powers between the state administration and academic institutions.[24]

One of the most important decisions affecting higher education is that related to admission quotas and selection criteria. Survey responses support the idea of giving a major role to the faculty in matters of content and organization of teaching, and a corresponding role to the chair with regard to research. It also recognizes the need for the participation of the academic senate, the administration of the institute, the institutions in charge of education and science, and other state agencies in matters related to governance and those affecting the whole academic community at the institutional or national levels. This applies in particular to the allocation of state funds to institutions of higher education, legal regulations concerning processes of appointments and the teaching loads of academic staff, monitoring the needs for academic staff and student training, and the general administration of student aid programs coming from the state budget. It is clear from the survey that while recognizing the positive role the ministry can play in policy-making and coordination, the Romanian academic community does not wish the Ministry to be an "enforcement agency," closely supervising the implementation of a common syllabus, standardized manuals, and the teaching process, or being involved in the everyday financial decisions of particular academic institutions.

While there is no "single voice" for the academic community, in Romania or elsewhere, the foregoing summary of the findings of the survey provides some significant pointers as to how the academic community in Romania perceives its future options and its traditionally liberal university-model preferences for the process of transformation in higher education and science.[25]

Concluding Remarks

It is often argued that the nation, the state, and its institutions need a "usable past" to validate the present and inspire faith in the future. In this regard, the situation of higher education in Romania is probably more difficult than that of other postcommunist democracies. As has been shown, the Romanian transition strategy needs to avoid many of the communist policies and "solutions" in education and science, because even those that represented a genuine attempt to introduce elements of modernity in general have been compromised by ideological and bureaucratic deformity. At the same time, the precommunist academic tradition of Romanian higher education is probably too distant to be of·direct use.

It is therefore still difficult to offer an overall assessment of the direction of the process of postcommunist transformations in Romanian higher education. Overall, the present situation in Romania could be described as the half-light between socialism and capitalism. Observers of the Romanian scene, while criticizing the ambiguous state of political and social affairs in the country, also point out that there are optimistic areas. The country has made considerable achievements in science, art, literature, and technological endeavors, but this may be mainly because "Romanians (as well as the ethnic minorities residing on Romanian soil) are a creative people with considerable resilience. They have endured repression and rapaciousness before, and they have emerged from the excesses of oppressive regimes with renewed energy."[26] It can be expected that these characteristics, together with the lessons of the past, will also find expression in the renewal of academic work in Romanian higher education and other academic institutions.

Notes

*The views expressed in this text are those of the author.

1. For an analysis of this model, see J. Sadlak, "Higher Education in Eastern Europe: The Evolution of the "Socialist" Model and Its Post-socialist Framework," *Higher Education Group Annual 1990*, 1 (1), 1991, pp. 117–40.
2. See statement made by the rector of Cluj University in N. Ulman, "Serving the State—In Romania, the Aim of Higher Education is to Aid the Economy," *Wall Street Journal*, March 21, 1974, pp. 9–10.
3. R.C. Bogdan, "Integration of Higher Education with Production and Research in Romania," *Higher Education in Europe*, 1 (2), 1976, p. 13.
4. I.I. Popescu and L. Vlasceanu, "Comrade Nicolae Ceausescu's Determining Role in the Establishment of a Developed Romanian Education System in Close Connection with the Current and Future Needs of the National Economy," *Revue de Pedagogie*, 1986, pp. 3–8.
5. In the 1960s the Soviet-inspired *aspirantura* form of postgraduate studies and the title of "candidate in science" was abandoned. They were replaced by the

"doctor" degree, which is awarded to a person who has completed the required preparatory stage, passed required exams, and defended a thesis; and the "doctor-docent in sciences" title (*doctor-docente in stiinte*) which is awarded to a person who already possesses a doctor's degree and who has achieved a substantial and recognized record in research or produced works of great value for the advancement of science, technology, or the fine arts.

6.   "Legea educatiei si îvatamîntului," *Buletinul Oficial al Republicii Socialiste Romania*, 14, December 27, 1978: 1–20. For its English translation, see *The Education and Instruction Act* (Bucharest: Editura didactica si pedagogica, 1978).

7.   D. Chitoran, *Higher Studies and Research in the Field of Economics in Romania—An Overview* (unpublished manuscript), 1991.

8.   S. Ghimpu, "Characteristic Traits of the Status of Higher Education Personnel in the Socialist Republic of Romania," *Higher Education in Europe*, 10 (2), 1985, pp. 28–36.

9.   See *Statut du personnel enseignant de la Republique Socialiste de Roumanie* (Bucharest: Editions didactiques et pedagogiques, 1969).

10. V.I. Lenin, "The Achievements and Difficulties of the Soviet Government," *Collected Works*, vol. 28 (Moscow: Progress Publishers, 1965), p. 70.

11. For N. Ceausescu's speech on this issue, see *Congresul al IX-lea al Partidului Comunist Romîn* (Bucharest: Editura politica, 1965).

12. Cincinalul 1976-1980—Cincinalul revolutiei technico-stiintifice, *Revista de Statistica*, 24 (2), 1976, pp. 54–60.

13. See chapter 6 of the report of the Council of Economic and Social Development, the State Planning Committee, and the Central Statistical Office in "Comunicat cu privire la îndeplinirea Planului National Unic de Dezvoltare Economico-sociala a Republicii Socialiste Rômania în perioda 1976–1980," *Revista de Statistica*, 31 (1), 1981, pp. 4–27.

14. I. Tripsa, "The Strategy of Romanian Scientific Research in the Eighties," *Journal of the National Commission of Romania for UNESCO*, 22 (1), 1980, pp. 123–26.

15. M. Florescu, "The Romanian School of Chemistry and Petrochemistry," *Journal of the National Commission of Romania for UNESCO*, 25 (2), 1983, pp. 243–49.

16. C. Bàlà, "University-Industry Relations: A Romanian Case Study," *Higher Education in Europe*, 8 (4), 1985, pp. 17–25.

17. J. Rupnik, "Higher Education and the Reform Process in Central and Eastern Europe," *European Journal of Education*, 27 (1, 2), 1992, pp. 145–51.

18. "Scientists Grapple with Eastern Mysteries," *Times Higher Education Supplement* (London), October 7, 1988, p. 11.

19. A separate Hungarian university functioned in Cluj until the end of the academic year 1957/58. For a history of this institution, see J. Sadlak, *Higher Education in Romania, 1860-1990: Between Academic Mission, Economic Demands and Political Control* (Buffalo: State University of New York at Buffalo—Graduate School of Education Publications, 1990).

20. See C. Woodard, "Universities in Former Eastern Bloc Seek More Links with the West," *Chronicle of Higher Education*, May 27, 1992, pp. A31, A33.

21. For example, financial assistance accorded by the European Communities within its PHARE program was (in millions of ECU, as of December 1992): Hungary 301, Poland 290, Bulgaria 115, Czechoslovakia 85, and Romania only 25.

22. *Lege privind acreditarea institutiilor de invatamint superior si recunoasterea diplomelor* (Bucharest: Parlamentul Romaniei, 1993).

23. See J. Hryniewica, B. Jalowiecki, and A. Mync, *The Brain Drain in Poland* (Warsaw: European Institute for Regional and Local Development of the University of Warsaw, 1992).

24. G. Vaideanu et al., *Academic Freedom and University Autonomy—Investigation Realized in Romania on the Occasion of the International Conference of University Teachers* (Bucharest: National Commission of Romania for UNESCO, 1992).

25. For a general analysis of this phenomenon, see Jan Sadlak, "In Search of the

'Post-communist' University: The Background and Scenario of the Transformation of Higher Education in Central and Eastern Europe" in K. Hüfner (ed.), *Higher Education Reform Processes in Central and Eastern Europe* (Frankfurt am Main: Peter Lang, 1995), pp. 43–62.

26.   T. Gilberg, *Nationalism and Communism in Romania: The Rise and Fall of Ceausescu's Personal Dictatorship* (Boulder: Westview Press, 1990), p. 276.

Chapter Seven
Czechoslovakia
Higher Education, Science, and Change

Vladislav Hancil

In 1992 it was decided that the Slovak and Czech parts of Czechoslovakia would become two fully independent countries. This paper principally emphasizes current changes and prospects in the Czech Republic.

Many new universities and scientific institutions were founded under the aegis of the Czechoslovak and Slovak academies of sciences (CSAS). In addition, individual ministries established numerous technological research institutes and laboratories; the number of their employees are comparable to those of developed European states. Unfortunately, this is not also true of their technological equipment, their databases, and their management. The originally high professional level of the workers was continuously lowered by the forced appointment of Communist Party members, very often without regard to their professional standing. Further, the standard was continuously lowered by the extensive emigration of scientists, a majority of whom found employment in foreign countries. Under the communist regime, science, technology, and the entire economy was directed toward total self-sufficiency in raw materials and industry, military preparedness, and exports to the COMECON (Council of Mutual Economic Assistance) countries and the Soviet Union.

The forced separation of Czechoslovakia from its traditional Western cultural, scientific, and industrial partners resulted in isolation and excessive specialization in science, and investment in uneconomical industrial state enterprises, dependent on economic links with the East. The low level of technology and the lack of coordination between science, technology, and production became a chronic problem for the centrally planned industry. These factors combined to create an economic structure that is only today slowly beginning to change.

Historical Development

Czechoslovakia's scientific activities are rooted in the past. They are connected with the establishment of Charles University in 1348, in Slovakia with the establishment of Academia Istropolitana in 1467, and, later, the University in Olomouc in 1573 and Trnava in 1635. The beginnings of technology originate with technical institutions such as Stavovska engineering school established in 1717, which had become a polytechnic school by the end of the nineteenth century.

The King Czech Science Society was established in the 1770s, then the only science society in the Hapsburg monarchy. The Czech Academy of Arts and Sciences was established in 1880 to promote scientific development and diffusion; the Czech area was then the most industrially developed part of Austro-Hungary.

After 1918, when Czechoslovakia was created, research and development was undertaken at universities and large industrial enterprises such as Skoda Plzen, Vitkovice Ironworks, Czecho-Moravian Kolben Danek, Bata, Poldi Kladno, and elsewhere. Small industrial enterprises rarely carried out their own research and development and tended to look to institutes of large enterprises and universities for assistance. Industrial enterprises made extensive use of licensing. It should be mentioned, however, that the research institutes of large enterprises dealt with both basic and applied science research problems, as did universities. No rigid separation of basic and applied science existed.

While industrial research was supported by industry, the state supported agricultural and forest research. In the 1920s, a complex of State Agricultural Research Institutes was created in Bratislava and Kosice and a Forest Research Institute in Banska Bystrica. Medical research was undertaken by universities and the State Institute for Health, established in 1925. Institutes in other fields were the Radiological Institute (1919), the Research Institute for Air Transport (1922), the Research Institute of Glass (1923), and many others. There were forty independent research institutes working in Czechoslovakia in 1937; 1.8 percent of the GNP was devoted to research and development.

Before the Second World War, there was no ministry or other office of the state responsible for science and technology. Many private businessmen realized the value of supporting research and development, and, after 1935, a tax concession was available to those who supported R&D and the fast depreciation of production machinery.

Czechoslovakia became a part of the Eastern bloc in 1948. The economy became centrally controlled as private property was elimi- nated, and that was reflected in the goals and direction of research administration. A new organization called the Research and Development Basis, which was a unified system of organizations and institutions, was tasked with supporting the Science and Technology Development Plan. In the year 1951, 14,000 people worked in the R&D Basis; in 1960, 92,000; in 1970, 147,000; and on December 31, 1989, 198,000. Administrative staff and support staff in R&D Basis represented about 43 percent of the total staff; in full-time equivalents it was even more.

Basic research has been concentrated mainly in the Czechoslovak and Slovak academies, which were similar to the Soviet model. The Czechoslovak Academy of Sciences was established in 1952 and the Slovak Academy in 1953 as a constituent part of the former until 1969, when they separated.[1] The 1968 Soviet invasion ended the process by which the Czechoslovak Academy looked after Czechoslovakia's international agreements for both the Czechoslovak and Slovak academies, rather than just its own institutes.

Universities ceased to be centers of basic research, and a combination of old equipment and poor motivation divided education from science. The academy was systematically excluded from the process of education. With some exceptions, scientists of the academic institutes were not allowed to lecture at universities. Under strict Communist Party control, however, they could educate new scientists and award them with the title "Candidate of Sciences" (CSc), which was generally accepted as equivalent to a Ph.D. Politically unacceptable persons were not allowed to stay at universities but were sometimes allowed to survive hidden in the institutes of the academies.

Applied research was, in the fifties, centrally controlled, and a network of research institutes was created that lacked effective links with production. Later the new system of control and restructuring of the "Production Technological Basis" attempted to link applied research units with production enterprises. But the approach was purely administrative, and its consequences strengthened the monopolistic production structure in Czechoslovakia. Equally unfortunate was the restructuring of the late eighties, which was done in hopes of combining positive elements of the planned and market economies. The changes led to a further decline in research quality because the key organization, the state research enterprise, which was owned by the state and self-financing, concentrated on short-term activities and goods and services that could be purchased easily outside the Eastern bloc. There was almost a complete absence of links between universities and the state research enterprises. Science and technology policy was formulated centrally; the main office was reorganized often and followed the dictates of the Central Committee of the Communist Party. There was little or no direct link with researchers or scientists.

Since November of 1989, when the Communist Party lost power, there have been many assessments of higher education, science, and technology in Czechoslovakia. Some of them follow: Stanford Research Institute,[2] Organization for Economic Cooperation and Development,[3] Centre Nationale de le Recherche Scientifique,[4] World Bank,[5] and most recently, Segal Quince Wicksteed, Ltd.,[6] Institute of Scientific Information,[7] and National Academy of Sciences.[8] Authors of individual studies have evaluated publication output and efficiency, impact factors, and other characteristics, or reported special features of the system in science and technology. All the reports found the publication output low, when related to the number of people involved in science, but not so low when compared with the financial investment in basic research. There was never any evaluation of teaching.

The totalitarian countries of the Soviet bloc developed huge academies, which even in the beginning were unable to compete with the West. Their efficiency diminished as a result of the Communist Party's wish to control everything. The rule *divide et impera* was applied and, in consequence, no education but only postgraduate training was allowed at the institutes of the academy. Given this situation, East bloc countries urgently need to formulate science policy and create a new structure. Yet too many scientists reject reorganization and stick to, and defend, the old hierarchical structures.

Higher Education

Reform of higher education's legal basis came quickly after the Velvet Revolution, in order to permit its implementation in the new 1990 school year. Act No. 172 on Institutes of Higher Learning and Universities passed the Czechoslovak Federal Assembly on May 4, 1990. The state, according to the law, is obliged to provide the major proportion of financial resources needed to operate the system, while the institutes and universities manage their affairs autonomously. There are today twenty-four institutions of higher education in the Czech Republic, of which fourteen are involved in science and technology. The numbers of students and teachers for selected institutions are found in Table 7.1.[9] Figure 7.1 is a diagrammatic representation of the structure of the overall education system and where universities fit into the hierarchial structure of the system relative to the age composition, on average, of the student population.

In 1991, roughly 16 percent of the twenty to twenty-four age group had access to the system of higher education. However, the overall size of the system is well below OECD levels. Half of the graduates from Czech academic and technical secondary schools find their way into higher education. Only forty-five to fifty percent of secondary CSFR enrollments are in the academic and technical secondary schools; it is these graduates who are eligible to compete for university places.

There appears to be little significant change in enrollments when compared with past reports. General universities, with their key faculties of science, medicine, humanities, and law, now have about 37 percent of enrollments; those in the technical or engineering universities are marginally greater. Economics universities have enrollments greater than 10 percent, a small proportion of the real need. There is an urgent problem of teacher quality, but this now has the assistance of such institutions as CERGE (Center for Economic Research and Graduate Education). Women represent slightly less than half of all enrollments; in pedagogical universities this portion was in the past significantly higher. Figure 7.2 shows fields of specialization of selected institutions of higher education.

Comparison of the quality of teaching staff is rather difficult because of the different system of academic degrees. Czechoslovakia was based on Soviet-type degrees, the CSc and DrSc. In some fields, a CSc degree can be considered a Ph.D. equivalent. Nevertheless, one-third of full professors and over 90 percent of associate professors do not have qualifications equivalent to a Western doctoral degree, and more than half of assistant professors have no formal academic qualification beyond their first degree. Many high-quality, independent people, who previously were not allowed to teach at universities, have yet to join the regular teaching staff at some universities.

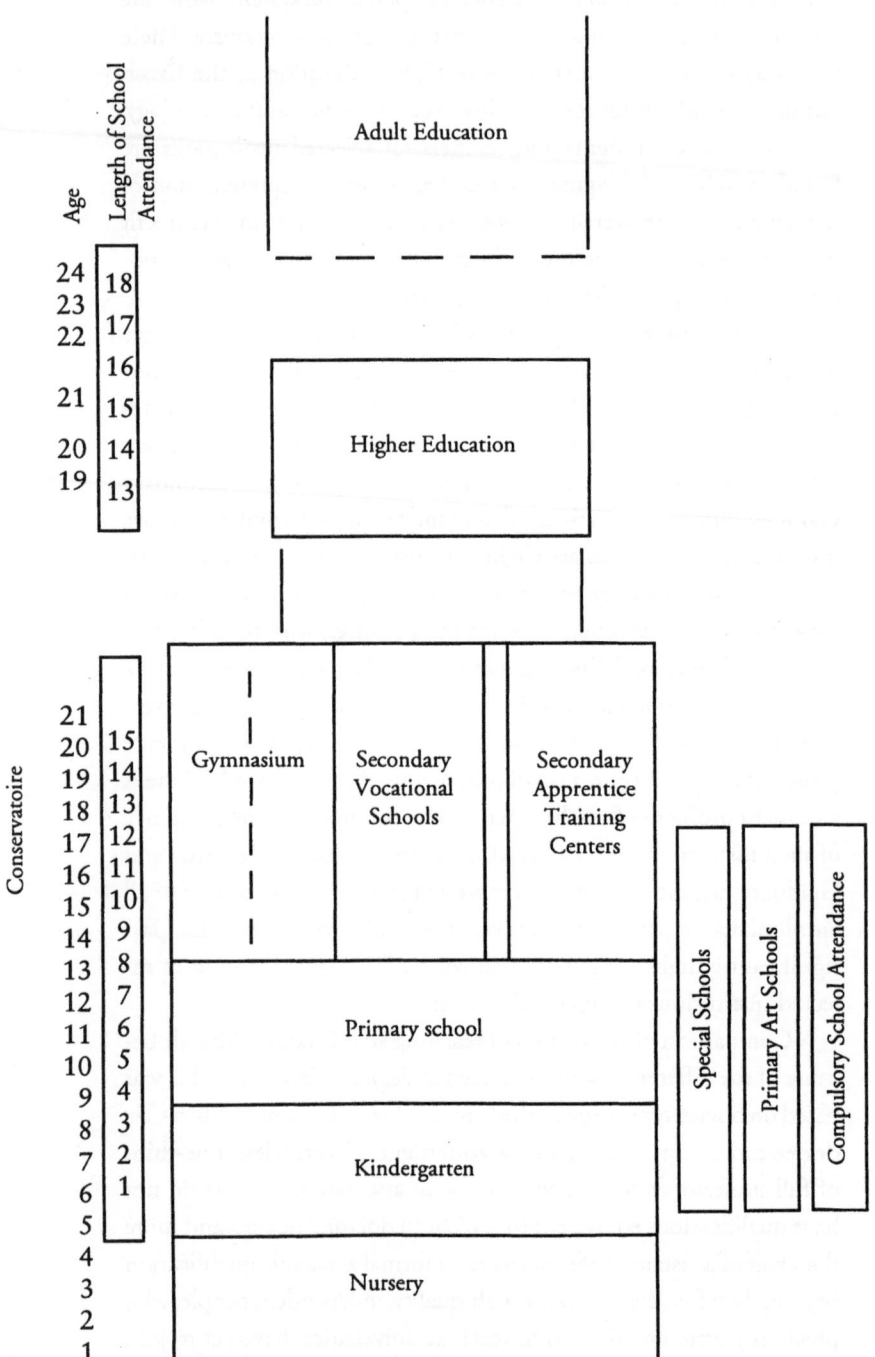

Figure 7.1 *System of Education in the CSFR*

Table 7.1 Universities Involved in Science and Technology in the Czech Republic

Universities	Faculty	No. of Students	No. of Teachers
Czech Technical	Mechanical Engineering	6906	466
	Electrical Engineering	4408	425
	Civil Engineering	755	412
	Nuclear and Physics	542	67
	Masaryk LAS	490	4
Technical (Brno)	Mechanical Engineering	3530	335
	Civil Engineering	3223	311
	Electrical Engineering	2572	281
	Technology	809	75
Technical (Ostrava)	Mining Geology	1595	170
	Mechanical Engineering	1387	128
	Metallurgy/Materials	1108	86
	Electrical Engineering	790	60
Charles	Sciences	1874	370
	Math and Physics	1708	216
	Pharmacy	887	90
Masaryk	Sciences	1970	214
Palacky	Sciences	1370	135
Chemical Technology	Food and Biochemistry	821	52
	Chemical Technology	817	127
	Chemical Engineering	470	145
	Environmental Protection	364	39
Chemical Technology (Pardubice)	Chemical Technology	858	126
Technical (Liberec)	Mechanical Engineering	1808	123
	Textile	1477	68
Western Bohemia	Mechanical Engineering	1350	177
	Applied Sciences	737	121
	Electrical Engineering	650	75
South Bohemia	Agriculture	779	101
Agriculture (Brno)	Agronomy	1165	155
	Forestry	596	89
	Mechanization	556	35
Agriculture (Prague)	Agronomy	1241	147
	Forestry	276	49
	Mechanization	763	85
Veterinary	Veterinary Medicine	721	80
	Veterinary Hygiene and Ecology	155	47
	Pharmacy	36	35

Higher education is financed from the budgets of the two republics. Students pay no fees and the universities charge nominal amounts for dormitory accommodations and food. Many students receive a stipend. Fees are charged to foreign students. Until now, institutions of higher education received a budget, and any independent income generated was to revert to the Ministry of Finance.

Expenditures by establishments of higher education in 1989 were 6.3 billion Kcs (Czechoslovak koruna), or approximately 17.5 percent of overall education expenditures and 0.8 percent of GDP. These expenditures have grown by

20 percent, nominally, in the last two years. Expenditures per student at universities in the Czech and Slovak republics differ significantly (see Harbison). Universities in the Czech Republic spent about 29,900 Kcs in the year 1989/90; in Slovakia it was about 36,800 Kcs. However, these figures may be misleading, as the differences in spending in individual universities vary widely from 12,000 (in economics universities) to 58,000 Kcs (in music and art academies).

Higher education in the CSFR is overstaffed when compared with many European countries; this is one of the main difficulties facing the educational system. Often, students have thirty hours of lectures a week and are unable to develop individually. In consequence, both the teaching staff and the students are overworked. Table 7.1 demonstrates the differences in this ratio. Act 172 on Institutions of Higher Education and Universities does not allow one to follow current trends statistically; this will make it more difficult in the future to analyze trends such as those described here.

Higher Education Act 172 gives the institution's relevant dean and faculty senate full authority to decide how many students to admit and how to select among applicants. This emphasis on academic judgments, a natural reaction to the situation prior to 1989, led to the unfortunate situation, in the year 1991/92, that some faculties could not accept more students because they had admitted too many previously. Each faculty sets its own entrance examinations and uses the results together with secondary school and teachers' recommendations as the criteria for acceptance.

Scientific Research

Since 1989, science policy in the Czech Republic has been dominated by the dispute between the academy and the universities. Centrally planned socialist regimes developed two sets of institutions that dealt with research. Scientific research was undertaken at the institutes of the academies of sciences, while applied research and technology was controlled by specific governmental departments or individual enterprises in their own institutes. There was very little research capacity within universities. The proponents of a new system argue that in all Western countries, research is carried out principally in universities and so, to enhance scientific research, science teaching and research need to occur side by side.

Figure 7.2 *Institutions of Higher Education and Fields of Specialization*

Name of Institution (Data from July 31, 1991)	Theology	Humanities	Law	Social Sciences	Economics	Pedagogy and Teacher Training	Natural Sciences	Agriculture and Forestry	Sports	Technology and Sciences of Technology	Medicine	Dentistry	Veterinary Medicine	Pharmacy	Architecture	Music	Performing and Dramatic Arts	Fine Arts and Design
Clazles University, Prague	x	x	x	x		x	x		x		x	x		x				
Mazzryx University, Brno		x	x	x		x	x				x	x						
Palmcky University, Olomouc	x	x	x	x		x	x		x		x	x						
University of Ostrava	x					x	x											
University of South Bohemia, C. Budejovice	x			x			x	x	x									
University of West Bohemia, Plzen						x	x			x								
J.E. Purkyne University, Usti nad Labem						x	x	x	x									
S. Lazian University, Olomouc	x		x			x												x
Faculty of Education, Hradac Kralove						x												
Czech Technical University, Prague		x								x					x			
Institute of Chemical Technology, Prague							x			x								
Technical University, Liberec				x						x								
Institute of Chemical Technology, Pardubice					x		x			x								
Technical University, Brno				x	x					x					x			
Technical University of Mining, Ostrava				x						x								
University of Economics, Prague				x	x	x												
University of Veterinary Sciences, Brno													x					
University of Agriculture, Prague							x	x	x									
University of Agriculture and Forestry, Brno							x	x	x									
Academy of Performing Arts, Prague																x	x	
Academy of Fine Arts, Prague															x			x
Academy of Applied Arts, Prague																		x
Janicak Academy of Music, Prague																x	x	
Comenius University, Bratislava	x	x	x	x		x	x		x		x	x		x				
University of P.J. Safark, Kosice	x	x	x			x	x				x	x						
College of Education, Zilna						x												
Faculty of Education, Banska Bystrica						x												
College of Veterinary Medicine, Kosice													x					
Siovax Technical University, Bratislava				x						x								
University of Transport, Zilna				x						x								
Technical University, Kosice				x						x								
School of Economics, Bratislava					x													
University of Agriculture, Zilna							x	x		x								
University of Forestry and Wood Technology, Zilna								x								x	x	
University of Arts, Bratislava																		x
Academy of Fine Arts and Design, Bratislava																		x

Table 7.2 Field Specializations of Research Institutes

Disciplines	Number of Institutes	Number of Researchers
Mathematics and physics	9	450
Technical science	10	230
Earth sciences	5	195
Chemical sciences	6	360
Molecular and cell biology	9	350
Ecological sciences	6	140
Medical sciences	8	210
Social sciences and economics	7	190
Humanities	17	360

Note: Considering the relatively large number of institutes, the average institute is small.

In the Czech (Slovak) Academy, which takes care of institutes in the Czech (Slovak) Republic only, there are eighty-two (sixty-one) formal research institutes. Some of them do not have a full range of administrative and support staff and so share these services. Each Republic has built its own independent capacity in most areas; this may not be appropriate, even following their split, because of their small size.

In the Czech Academy of Sciences (Czech Republic), the division by disciplines is shown in Table 7.2; a similar division is found in the Slovak Academy.

Table 7.3 Professional Size of Research Institutes

Number of Scientists	Institutes (Social Science Excluded)
1 to 25	23
26 to 50	20
51 to 100	6
more than 100	4
Total	33
Percentage with fewer than 50	80%

According to available statistics, CSFR spent 1.2 percent of its GNP on research in 1988, well below the average 2.7 percent share in many OECD countries. The proportion devoted to basic science is equally reduced. Only 10 percent of research funding (nearly zero for basic science) comes from industry.

The total number of publications per capita was analyzed by Welljams-Dorof for the period 1981–85 versus 1986–90. He found a

Figure 7.3 *Structure and Function of Grant Agency of the CSAS*

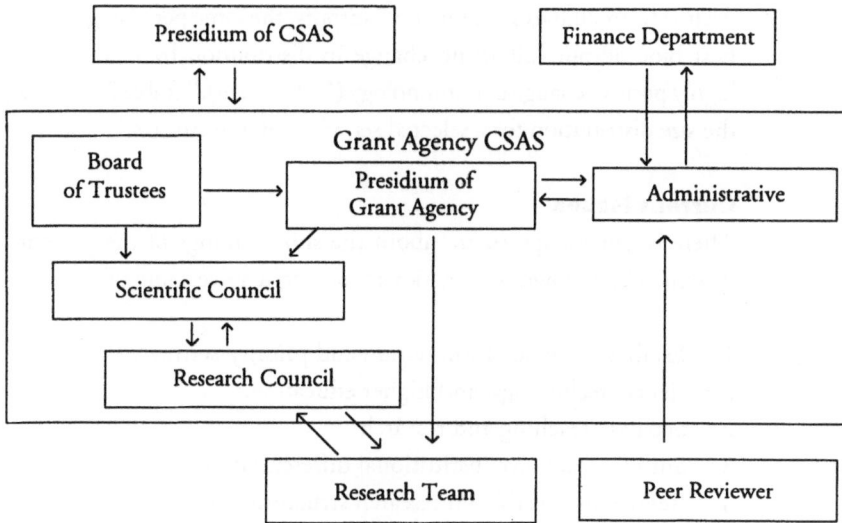

Notes

Board of Trustees: Body of trustees in which the representative of the other organizations important in the science and technology system are delegated.

Presidium of GA: Four members, a president, and three vice presidents. The vice presidents are responsible for the following scientific divisions: physical and technical sciences; biological and chemical sciences; economics, social sciences, and humanities.

Administrative: Responsible for all administrative work and controlled by the presidium of GA. It is also an information centrum on running research projects in the academy, and it maintains the corresponding databases.

Finance Department: Responsible for a part of the bill of the CSAS that was assigned to the Grant Agency. The part is divided into accounts of individual research councils and budget of the office of the Grant Agency.

Scientific Council: Coordinates the activities of individual research councils and decides on all quarrels, which are related to the evaluation of the scientific quality of grant applications. The members of the presidium can take a part in the negotiations of the scientific council. Chairmen of research councils are members of the scientific council, and the same number of scientists can be elected directly by the science community of the CSAS.

Research Council: Nine research councils were established: mathematical and physical sciences; technical sciences; earth sciences and astronomy; chemistry; medicine; ecological sciences; economics and social sciences; humanities; history and philosophy. The Research Council has twenty members who are elected by the science community; they elect a chairman from the members of the Council.

Research Team: A scientific group or an individual from the institute of the academy.

Peer Reviewer: Unpaid reviewer evaluating the grant application. The numbers of reviewers of an individual grant is proportional to the financial support requested by the applicant. There should be at least three peer reviewers and one of them should be from abroad. The peer reviewers are chosen by the research council from a databank that is continuously updated.

Councils: Councils are elected; the elections are open to the elected scientific academy employees. Votes are delivered by mail from the list of candidates, which is prepared on the basis of suggestions from individual institutes. The number of candidates can be limited by the presidium of the grant agency.

decline from 20.8 to 18.8 for cited outputs; the same percentage from East bloc output (20.3); no change in distribution by field; and a highly positive change in immunology (102 percent).[10] Table 7.3 shows the size distribution for a selected set of research institutes.

Current Issues

There is general agreement about the shortcomings of the present system. The following are actions that might solve some of them:

1.   Establish a policy-formulation and priority-setting mechanism for science, technology, and higher education.
2.   Separate teaching and research.
3.   Simplify excessive institutional differentiation.
4.   Reduce the overblown research structure.
5.   Increase overall size and provide greater program options in higher education.
6.   Identify major gaps in the curriculum of higher education and the content of research.
7.   Improve the uneven quality of academic and scientific personnel in the context of generous overall staffing levels.
8.   Alter the inequitable and inefficient system of student selection to higher education.
9.   Improve inefficient resource allocation and use.
10. Vary single-source financing from the public budget.
11. Develop international links to counter the accumulated impact of forty years of deprivation of international peer interaction, and permit access to Western literature and modern equipment.

A World Bank study by Harbison states:

No country has unlimited resources for development of higher education, science, and technology. A system for establishing broad priorities, in line with the likely development path of the country, has everywhere proved to be essential and more so in relatively small countries which must husband resources and choose areas of concentration with particular care. CSFR appears to have no such system at present. The old style closed structures of central bureaucratic command responsive only to Communist Party direction have been eliminated. But nothing has yet

Figure 7.4 *Scheme of the Supposed Grant Agency of the Government of the Czech Republic*

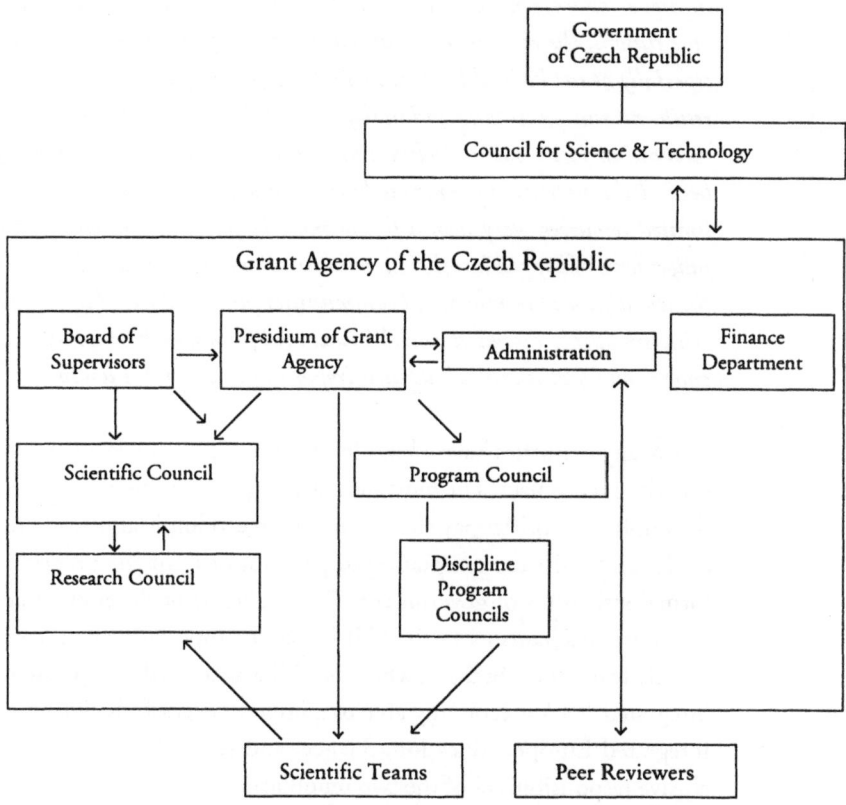

Notes

Presidium GA: According to law there are a president and four vice presidents. Vice presidents are responsible for scientific section; priority program section; international relations; and finance. The presidium determines which portion of the budget assigned to the individual research councils will be distributed, with the highest priority given to joint programs. The joint programs involve the cooperation of existing institutions and they will help to overcome past barriers.

Administrative/Finance: An information center of GA and of the council for science and technology.

Scientific Council: Like the NSF.

Program Council: Coordinates activities of individual program councils.

Research Councils: There are ten research councils: mathematics and physics; technical sciences; earth sciences and astronomy; chemistry; biology; medicine; ecology and environmental science; agriculture; social sciences and economics; humanities, history, and philosophy.

Discipline Program Councils: There are seven discipline councils: telecommunications and information; production technology and material sciences; agriculture; environment and energy; health protection; social problems; education.

emerged in their place to provide broad orientation to the nation's efforts in higher education, science and technology. For entirely understandable reasons, the mere mention of coordination of policy and programs, especially at the Federal level, evokes a knee-jerk negative reaction. As a result, necessary thinking and dialogue on the evolution of a more open mechanism to set overall policy and priorities for the sector is lagging behind the necessity to confront hard decisions in the context of very limited resources. Very soon a forum must be found in which the best judgements of key public officials, of business managers, and of leaders of the scientific and technological communities can be subjected to mutual criticism openly discussed and debated, and from which consensus can emerge on what is critical, what is important and what is less so.[11]

Many countries have a high-level council or other form of consultative mechanism. The most successful experiences are those where the public authorities pay close attention to economic and academic circles and resist the temptation to preempt or limit the process to formal structures of government. The challenge of developing appropriate mechanisms in the CSFR may be especially great—and especially urgent—because, while the CSFR seeks to develop a single integrated market economy able to compete successfully within an integrated Europe, education, science, and technology are the exclusive responsibilities of the two republican governments. Both assiduously guard their prerogatives. However loose and even informal it may be, some form of coordination at the federal level is essential if unnecessary duplication and waste are to be avoided.[12]

The following reforms have been undertaken:

1.   Universities established the important Council of Universities.
2.   Senior-level staffing at the CSAS was changed considerably, with 87 percent of the institute directors being replaced. The members of the presidium (the central steering body of the academy) are newly elected. As a result of these changes, together with economic pressure on the academy, the CSAS has:

   •   formed a Coordination Council of the Czech Government for Science and Technology as of the beginning of this year;
   •   taken an active part in preparation of Act 300 of the Czech

Figure 7.5 *Allocation of Funding*

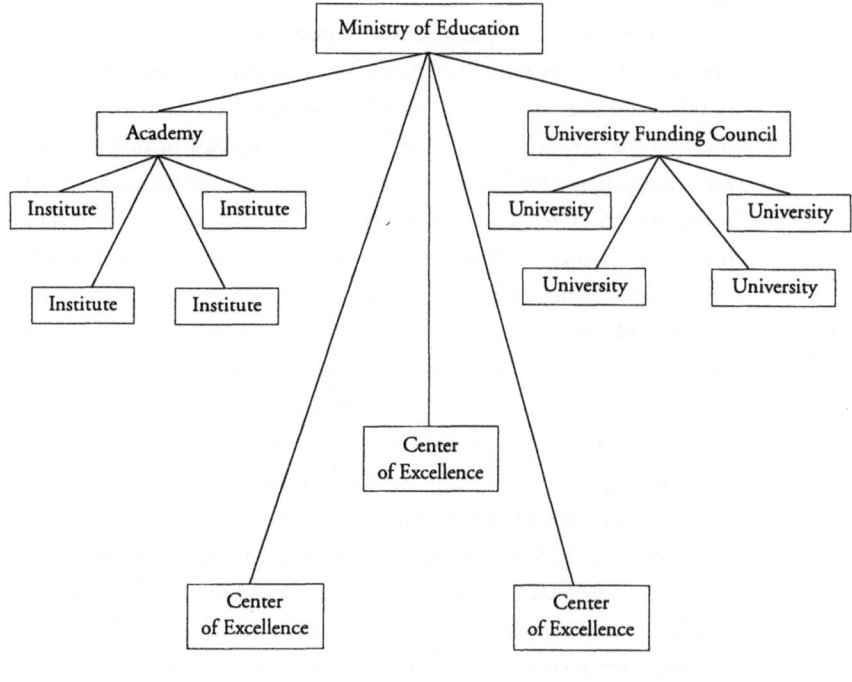

Parliament for state support for science and technology;
- prepared a new act for the academy of Sciences of the Czech Republic (Act 283/1992);
- established its Council for Conception, which was originally intended to set overall priorities and directions for scientific research inside the academy; its role changed to determination of policy on the transformation of the academy, and the activity of this council has now however been suspended;
- established an internal grant agency to introduce grants rather than institutionally based funding.

The Internal Grant Agency of the CSAS was the first one in Czechoslovakia. The idea was also applied in the Ministry of Health. Figure 7.3 says more about the Agency's function. The Act on State

Support for Science Activity and Technological Development was accepted by Parliament on May 4, 1992; the structure of the Grant agency is shown in Figure 7.4.

The newly elected parliament and government do not agree completely with the structure just described, objecting to the existence of the Council of Science and Technology, which should be the advisory body to the government in this field. The criticism principally calls for a dual scheme for financing the academy and universities as shown in Figure 7.5. The Centers of Excellence have not had an important positive influence until now.[13] It is quite possible that there will be two grant agencies in the future, both derived from the budget of the Ministry of Education, and that resource allocation by grant competition will be partly substituted for by issuing orders. This process will depend largely on the ability of the administrative staff to choose wisely. The prospects for this are not very encouraging at the moment.

Although there is strong political pressure otherwise, there are strong arguments for the Council for Science and Technology and its proponents to look for greater efficiency and more effective links to government. Ministry grant agencies should be examined, which will lead inevitably to structural changes. It will also help formulate priorities for science, technology, and higher education; reduce the separation of teaching and research personally and institutionally; formulate better combined programs between higher education and research; improve the quality of academic and scientific personnel; and make administration simpler.

Building the structure has only just commenced. Nonetheless, a number of conclusions can be reached, as already noted:

1. Policy has not been formulated and the priority-setting mechanism has only a legislative base—that is, it has not been accepted by the Czech government.
2. The separation of teaching and research continues.
3. Institutional differentiation continues.
4. The overblown research structure has been significantly and drastically cut. However research mobility, particularly abroad, hurts intrinsic Czech interests.
5. Very little has been done with regard to economies of scale because of limited resources and the low absorptive capacity of institu-

tions of higher education; there are no common grant programs.

6. Little has been done to lessen the gaps in the curriculum of higher education and research themes; the Ph.D. programs of the institutes of the academy provide much of the research content.

7. Uneven quality of academic and scientific personnel, together with generous overall staffing levels, significantly changed when the academic institutes reduced personnel by 30 percent.

8. Individual departments' autonomous behavior has resulted in much higher efficiency and less inequality in student selection, but it also increased differences between individual departments.

9. A local peer review system was used for resource allocation.

10. Industry cannot find resources for financing basic research and has no resources to finance applied research.

11. The deprivation of international peer interaction is being strangely compensated. The low mobility of scientists in the past is being replaced by a unidirectional flow of scientists from Czechoslovakia, compensated by a flux of science policy analysts from the West. The price increase for books and journals has not allowed for an improved literature flow. The communications network, because of its quality, will not permit a wide use of computer technology. The uneven availability and application of modern computer technology sometimes reduces real research activity to a useless study of obsolete computer programs.

The intention of the government, stated in its program for 1995, defines the growth of support for science from 0.46 percent to 0.7 percent of GNP. However, the available financing of the academy has stagnated at the 1994 level. Even the inflation is not compensated. Salaries are 8.4 percent higher (half of the rate of inflation). In the years 1989 to 1993 the academy reduced its staff by 50 percent.

In 1994 another 500 people (of about 5,000 at the beginning of 1994) left the academy, which was not intended. The main reason for this was that salaries outside the academy can be easily three times higher. Those that left the academy were young, prospective scientists. They also left because of the problem of aging equipment. Funds for replacement of equipment currently represent only 3 percent of the total academy budget.

Unfortunately, even the new grant agencies have slowly become a support for the status quo. There are now many grant agencies in the Czech Republic, e.g. GA of Czech Republic, GAB of Academy, GA of Czech Technical University, GA of Ministry of Health, and so on. All of them are trying to support individuals and groups of workers from their own institutions. Cooperation between institutions is either poor or nonexistent.

On the other hand there are some early signs that the attitude of people toward science may be changing. For example, the only regularly broadcast scientific program on Czech radio is the most popular program of all. Hopefully, these changing attitudes may signal a change in government priorities and the infusion of the necessary funding to permit the process of real change to occur.

Notes

1. The Czechoslovak Academy of Sciences was the successor of the Czech Academy of Sciences, established one hundred years ago and following from its beginning the ideas of the great Czech scientist J.E. Purkyne, who anticipated and, in fact, initiated the formation of the Czech Academy of Sciences. It not only gathered prominent scientists, but also ran several scientific institutions.

2. Catherine P. Allex, H. Robert Coward, and Ronald R. Fresne, "Research Activity in the Czech and Slovak Federal Republic: A Bibliometric Assessment Overview," Stanford Research Institute: Report for NSF, March 1992.

3. Review of National Science and Technology Policy, OECD *Examiners Report* (Paris: OECD, May 1992).

4. Pierre Volfin, *Report* for the European Mission in Czechoslovakia (Prague-Bratislava: CNRS, March 30–April 4, 1992).

5. Ralph Harbison, "Notes on Higher Education and Scientific Research," *Social Impacts of Structural Adjustments* (Washington: World Bank, February 1991).

6. *Strategy for Science and Technology Policy in the Czech Republic* (Segal, Quince, Wicksteed, Limited, October 1992).

7. A. Welljams-Dorof, speech entitled "A Citationist Perspective on Czechoslovak Science 1981–1990," presented at the 16th World Congress of the Czechoslovak Society of Arts and Sciences, June 1992. (A. Welljams-Dorof is from the Institute for Scientific Information, U.S.A., Prague, Czechoslovakia.)

8. *Science, Technology and Economic Growth, The Case of Eastern Europe*, Summary of a Meeting (Washington: National Academy Press, 1992).

9. Figures 7.1 and 7.2 are presented in Maria Hrabinska et al., *Higher Education in the Czech and Slovak Federal Republic, Guide for Foreign Students* (Bratislava: Slovak Educational Publisher, 1992). Figure 7.1 does not include the system of academic institutes and some new universities, such as Trnava University.

10. Welljams-Dorof, *op. cit.*

11. Harbison, *op. cit.*

12. Federalism is now dead. The majority of citizens in Slovakia elected people who wanted a sovereign Slovakia. The majority in the Czech Republic voted for fast reform, which they believe should not be slowed down by political problems in the other part of Czechoslovakia. So the Czech and Slovak republics have organized separate systems.

13. It may be possible that the experience of the author led him not only to judge what is the best, but also what is feasible under current political constraints in Czechoslovakia.

Chapter Eight
Germany
Restructuring of the Universities, Colleges, and Research Institutions in Eastern Germany

Volker Lenhart and Stephan Stockmann

The political changes in the German Democratic Republic (GDR) since the fall of 1989, particularly reunification, have subjected institutions of higher education and research in the New Federal States (NFS) and East Berlin to a far-reaching process of restructuring. The first attempts at reform were aimed primarily at internal democratization together with a change of teaching and research content directed toward a convergence of the educational systems of the Federal Republic of Germany (FRG) and the GDR. These steps were interrupted by the rapid introduction of national unity and were superseded by restructuring stipulated in the reunification contract. These structural changes affect three areas: the law of higher education; forms of higher education and research institutionalization; and personnel and their renewal in the institutes of higher education and research.

Unlike institutions in other countries in Central and Eastern Europe, East German academic institutions, because of their integration into the economic and political system of the FRG, have had access to Western FRG's economic and personnel resources. This facilitates restructuring but also brings with it specific problems, such as competition between graduates and scholars from the two different systems of higher education and research. The following pages focus on changes that occurred during the process of merging these two societal systems and the resulting repercussions. Table 8.1 shows expenditures on education in the years from 1980 to 1991.

Situation of Higher Education and Research before 1989

The system of higher education and research of the GDR was characterized by government planning within the economy and a society-wide planning framework. There was a distinction between teaching and learning in institutions of higher education on the one hand and research on the other, the latter being mostly concentrated in institutions outside of universities.

In the GDR there were 54 institutions of higher education (among them 6 traditional universities) with 132,423 matriculated students in 1988. The number of new matriculations was about 32,000 per year, which represented a share of 13.7 percent of each age group (compared with 25.4 percent in the FRG). The number of students remained constant or even decreased during the mid 70s. This was due to strongly regulated access to higher education based on the government's view of future demand. The proportion of diploma- and other degree-holders among employees was 15.9 percent (1.23 million) in 1985, compared with 10.5 percent (2.57 million) in West Germany. Nevertheless, the planning process did not work very well, and many higher education graduates were employed in jobs not equivalent to their qualifications. The system of higher education was completely financed by government; the expenditure share for institutions of higher education was 1.4 percent of the national budget in 1988, or 24.2 percent of all educational expenditures.

Though the tasks and functions of institutions of higher education in the GDR retained the German tradition of unity of research and teaching, university research was severely truncated when compared with that of West German universities. East German universities were responsible for continuing education and the education of the "socialist personality." Their subordination to party goals and isolation from international trends led social sciences, humanities, economics, and law to problematic separate developments.

The pattern of research and development allocations between industrial research, research in institutions of higher education, and research in specific research institutions outside of universities was similar to that in many developed, industrialized nations. Trade and industry accounted for 77 percent of all financial resources and 64.8 percent of scientific personnel. The relatively large degree of governmental support provided for research and development outside of the university structure and for the institutes of the Academy of Sciences is remarkable. The academy administered basic and applied

Table 8.1 Expenditure on Education as Classified in the Education Budget in Territory A (West Germany)

Key	1980	1985	1988	1990	1991
DM Billion					
Preschool education	3.5	4.3	5.3	6.0	7.9
Grammar and Secondary Schools	43.8	46.8	49.7	52.7	55.8
Institutions of higher education	17.8	21.4	25.4	30.3	30.7
Continuing education	2.3	3.0	2.9	3.2	3.8
Promotion measures	6.2	5.3	4.6	5.2	7.1
Training assistance	3.6	2.3	2.4	2.7	4.0
Joint research promotion by the federal government and the Laender	3.6	4.5	5.2	5.4	5.8
Total	77.1	85.3	93.1	102.8	111.1
Percentage borne by the federal government	8.8	8.1	7.5	7.9	10.7
Laender	72.4	75.1	75.8	75.8	73.4
Local authorities	18.9	16.8	16.7	16.2	15.9
Expenditure on staff	56.2	60.1	60.7	61.0	58.1
Expenditure on material and equipment	16.0	18.8	18.0	17.7	18.8
Investment in fixed assets	12.6	7.8	8.1	7.8	7.4
As percentage of the overall public budget					
Total	15.1	14.1	13.9	13.8	—
Schools	8.6	7.7	7.4	7.1	—
Institutions of higher education	3.5	3.5	3.8	4.1	—
As percentage of the national product					
Total	5.2	4.6	4.4	4.2	4.2
Schools	3.0	2.6	2.4	2.2	2.1
Institutions of higher education	1.20	1.17	1.21	1.24	1.16
DM per Inhabitant					
Total	1,251	1,397	1,509	1,613	1,722
Schools	710	767	806	826	864
Institutions of higher education	288	351	412	476	476

research in mathematics, sciences, medicine, engineering, social sciences, and the humanities. The ratio of university research and development personnel, when compared with the academy's institutes, was 1:2.3 in 1987 (FRG 1:0.8). Research and development staff, amounting to 140,000 employees, represented 16.7 of each 1,000 gainfully employed persons in 1989 (FRG 16.8). This record, however, must be qualified for performance, and the fact that the financial conditions for research and development in the GDR were far worse than those in the FRG,[1] as a consequence of the lack of foreign currency and embargoes.

In the Academy of Sciences:

1.   Research was not autonomous but depended on central management, which itself depended on economic and social development plans.

2.   The research institutes were entitled to award doctoral (and postdoctoral) degrees without reference to universities. Universities were also disconnected from nonacademy-based research institutes, such as the Central Institute of Higher Education at Berlin, which was placed under the authority of the Ministry of Higher Education when it was formed in 1982. The institute became a research branch of the ministry. Over time, the academy's institutes did more and more directed research for government.

3.   Innovations were hampered by concentrating complete research branches at central institutes. The scope of basic research was narrowed. There were so-called achievement contracts between industrial enterprises and the academy's research institutions, and product development was often transferred to the institutes.

4.   The institutes of the Academy of Sciences increasingly became organizations for different economic sectors, such as trade and industry.

5.   International relations were limited to COMECON nations and developing countries. Scientific relations were impossible with Western nations in certain sensitive areas.

Reform Efforts, November 1989 to October 1990

Institutions of higher education and research played only a minor role in the radical political changes driven by the reform and democracy movement. The democratic renewal of institutions of higher education and the Academy of Sciences resulted mainly from the initiatives of the employees and students, not their administrations. New democratic regulations were implemented only slowly and became obsolete with German reunification after June of 1990. The Ministerial Council of the GDR agreed to the democratic administration of an autonomous Academy of Sciences before the Academic Council of the FRG was put in charge of its evaluation. The reunification treaty, which liquidated the Academy of Sciences, brought this development to an end as well.

Similar developments occurred at the institutions of higher education. Many regulations developed by the individual institutions as part of the process of internal democratization were subsequently replaced by state laws.

This period was also characterized by personnel changes in the administrative committees, the first liquidations of institutions (principally those that were most ideologically biased), and curriculum reform. After these initial changes, FRG structures were largely adopted.

Federalism and Academic Institutions
Contrary to the highly centralized system of the former GDR, education has a distinctly federal character in the FRG. Each state decides legislation for institutions of higher education and research. Federal cooperation is restricted to the HRG (*Hochschulrahmengesetz*, the framework law for higher education) and the common task of building and expanding institutions of higher education (Constitutional Law, Article 91a). In the area of research outside of universities (Constitutional Law Article 91b), the federation and the federal states can support common research institutions of supraregional importance. The Federation and State Committee for Educational Planning and Research Enhancement (Bund-Länder-Kommision) serves as a coordination agency.

Regulations of the Reunification Treaty In the reunification of August 31, 1990, regulations were introduced regarding the integration of institutions of higher education and research of the GDR into the reunified Germany. Article 38 (academia and research) of

the reunification treaty regulates the renewal of universities and research. By December 31, 1991, all institutions were to have been
evaluated by the Academic Council. Article 38, paragraphs 2 and 3,
provide for the liquidation of the Academy of Sciences, with individual institutions assigned to the federal states and employment
termination on December 31, 1991.

The transfer of authority to the federal states over institutions
of higher education is found in Article 13 (transition of institutions).
The state governments consequently regulate integration, or
Abwicklung (liquidation). *Abwicklung* is a German juridical term
meaning "the proper ending of the activities of an institution." State
parliaments were given until October 3, 1993, to vote on their own
higher-education laws, consistent with the federal framework law
for higher education.

For higher education, the Academic Council issued a statement
about possible future structures. Adoption will be regulated by the
individual federal states, which have developed evaluation procedures for those individuals currently employed in institutions of
higher education. In the area of research, study groups and institutes affiliated with the Academy of Sciences were evaluated by committees of the Academic Council. In these cases, individual performance was less of a priority than the quality of research as judged by
national and international standards.

According to Article 13 paragraph 1.4 of the reunification treaty,
state governments were obliged to decide if institutes were to be
integrated or liquidated. Those liquidated were principally specialized institutions of higher education and university departments for
Marxism-Leninism, law, economic studies, education, psychology,
and some areas of the humanities. In many cases, however, classes
were continued by an affiliation with other institutions of higher
education, the help of guest teachers, or the provision of limited
employment contracts, so that studies were rarely interrupted.

State Structural Committees were assigned the task of advising
the state government based on the framework established by the
Academic Council. Their tasks include: (1) advising on the founding of new institutions of higher education, especially the founding
of universities and specialized colleges (*Fachhochschulen*); (2) advising on curricula, particularly the establishment of new subjects and

revision of existing subject content; (3) advising on larger investments in institutions of higher education and research institutions; and (4) advising on the establishment of appointment committees.

The members of the Higher Education Structural Committee are appointed by state governments after consultation with the relevant state Rector's Conference. There may be no more than ten to twelve members. The majority of members should be acknowledged scholars with a high academic reputation; in addition, public personalities associated with academia may be appointed. Members do not come from the federal state concerned.[2]

The State Structural Committees began work in 1990 and their recommendations formed the basis for decisions of the state governments. Their recommendations took account of the guidelines of the Academic Council which, between March of 1991 and May of 1992, proposed a future structure for the system of higher education in the NFS. The recommendations contain detailed suggestions for the establishment and expansion of institutions of higher education as well as their distribution between federal states, taking into account estimates of the number of students in particular subjects. These estimates are based on intake capacity, a measure commonly used in the old FRG.

State Legislatures The individual states' legislation followed the reunification treaty in requiring laws for higher education to be adapted to those of the FRG by October 3, 1993. Until then, institutions of higher education were governed by the law of October 8, 1990, enacted by the last GDR government shortly before reunification. The law already included provision for formal adjustment to FRG structures. The federal states also passed laws to permit restructuring, formation of committees, and so forth. Higher education is now being regulated according to these laws, based on the recommendations of the State Higher Education Structural Committees and the Academic Council.

Personnel Validation A key issue for institutions of higher education was the renewal of personnel; dealing with this problem was regarded as a precondition for providing institutions with full autonomy. The evaluation process usually consisted of a three-step procedure, which varied slightly by state: past behavior (honor procedure), transition, and acceptance.

The honor procedure evaluates the behavior of full-time academic employees (in some states, of all employees) during the SED government. Institutions of higher education or the ministries established honor committees, based on agreed criteria, to decide on the classification of investigated persons according to the degree of their incrimination, that is, the extent of their active collaboration with the previous system. The honor committee classification decided how far and under what conditions the persons concerned entered the next phase of evaluation.

The professional qualifications of teachers and scholars were examined in the transition procedure. The transition committees at the institutions of higher education received the publications of all applicants, which were then judged by previously appointed experts. Subsequently, the relevant authorities, for example in the ministries, decided on the transition and on the recommendation of the committees.

The acceptance procedure was designed to assist with the reappointment of validated personnel for positions within the new structure. A certain number of professorial posts were filled directly, while others were advertised and then filled by regular appointment procedures. As far as other academic and technical personnel are concerned, committees at the universities and their departments decided, after evaluation of the applicants' professional qualifications, who would be reconfirmed in a position. Personnel not accepted were laid off according to the regulations of the reunification treaty. This step involved reducing academic personnel by about 20 percent (from about 22,500 to 16,500) by the end of 1992. In newly founded schools and institutions of higher education (newly founded in some cases after the old institution was liquidated), positions were filled by regular appointment and employment procedures. For appointments to advertised departmental positions, the appointments committee, however, is selected according to the recommendations of the Academic Council.

Research In Article 38 of the reunification treaty, the Academic Council was authorized to evaluate the institutes of the Academy of Sciences. The last democratically elected GDR government decided that from among the smaller academies not belonging to the Academy of Sciences, the Academy of Pedagogical Sciences would not be evaluated. The overall goal is for research to integrate science and research into FRG structures. East German reconstruction has been pursued in two directions: first, the sectoral and disciplinary expansion of FRG research structures outside of universities to the NFS; and second, the development of state regional policies for higher education and research. The supporting role of research outside universities is typically seen as follows: "In a federal system the federation and the states should see their role in the area of research outside universities primarily as additional funding of basic research on topics which will not or not yet be touched by universities. Further, their national role is taking care of commissioned research for ministries and industry."[3]

In the FRG, research external to the universities but supported by private industry is undertaken by four types of institutions. Excluding the Fraunhofer Society, they are financed completely or largely by the state:

1. The Max Planck Society (MPG) concentrates on basic research and supports new research trends that have yet to find any adequate place at the universities due to their mainly interdisciplinary character or expense. The MPG is a public research organization, an association formed under civil law, and is 80 percent financed by the federal government and individual states.[4]

2. The Fraunhofer Society (FhG) is practically oriented and is principally financed by contract research. It works on research assignments for governmental departments, political counseling, some aspects of basic research, research and development for industry, technology transfer to small- and medium-sized enterprises, staff recruitment for industry, and so forth. The FhG Institutes are financed in equal portions by available basic funds from the federation and the project funds of public organizations and industry. The average rate of self-support—that is, funds from projects and industry—is around 80 percent for each institute.

3. Large research institutions (GFE), such as the German Cancer Research Center in Heidelberg, represent another category. They work on research and development projects in the field of natural

sciences, technology, and medicine requiring interdisciplinary cooperation and providing economies of scale in equipment and personnel. The Federal Ministry of Research and Technology generally covers 90 percent of the funds and the individual federal state, where each institution is located, the remaining 10 percent.

4. Institutions on the "Blue List," which jointly finance (federation: state/50:50) smaller projects from different fields of science.

The distribution of the budget and personnel among the four groups in 1988/89 was as follows:

	Finance 1988 (in millions of DM)	Personnel 1989
MPG	1,084	8,718
FhG	625	3,843
GFE	3,323	21,379
Blue List	467	5,155

East German research institutions outside the university were not evaluated by individuals but as institutions. Goals involving sectoral and disciplinary expansion on one side, and regionalization on the other, were part of the basis of evaluation. In addition, a restructuring of nonuniversity research that involved a proportional shift to these four groups proceeded parallel to these evaluations and reinforced the new research policy.

The institutional evaluation was carried out by surveys and on-site visits. Evaluation criteria included citation indices, external financial support, the proportion of research and development to service activities, and so forth. Institutions not evaluated positively were liquidated—that is, units were dissolved and the personnel dismissed. Positively rated work areas were integrated into the new system. Previous structures developed over decades were rarely maintained. Instead, the process usually resulted in a redistribution of the professional personnel to newly established study groups and institutes under a new administration (mostly West German scholars). Thus, the continuation of the positively rated work areas took place in a much changed work context.

The recommendations of the Academic Council show a shift from the relatively independent institutions of the MPG and GFE

toward the institutes of the Blue List and FhG. New Blue List institutions need to obtain 50 percent of their financing from external
project support and so very much rely on the supply of those external projects. Thus, the Federal Ministry of Research and Technology has a greater responsibility for exercising control over research
groups and can now, as a large project investor, intervene further in
research development.[5]

*During the distribution of these posts to the various research sectors out-
side the university, an apparent restructuring has taken place. The Academic Council recommended, for instance, considerably fewer posts for
the large research institutions in the new states than would have been
expected, considering the entire size of the population and the research
capacity. An over-proportional emphasis on promoting the Fraunhofer-
Institutes and especially the institutions on the Blue List is recommended
which was, nonetheless, determined by the strong orientation of the institutions towards practically oriented and applied research and development as well as by their counselling and service performance. In the
unified Germany, the Blue List includes around 75 to 80 institutions
with more than 8500 personnel. Thus, it reaches a point where planning . . . for the development of research needs cannot be delayed, as is
already the case for the MPG, FhG, or other large research institutions.
It is, therefore, urgent to work on a concept for reshaping the Blue List.
Thus, the question of the current evaluation and the restructuring of
each institution should especially be considered.[6]*

In 1992 the distribution of finance and personnel among the four
groups was as follows:

	Budget 1992 (millions of DM)		Personnel 1992
MPG	1,314		10,350
Fhg	821	(1991)	4,900
GFE	2,864	(1993)	23,500
Blue List	1,000		9,000

Overall, the following institutions were established according
to the recommendations of the Academic Council: two institutes as
well as twenty-eight study groups of the MPG; nine institutes as

well as twelve branch offices of the FhG; three large research institutions and eight branch offices in the GFE; and thirty-four new Blue List institutions.

The greater number of Blue List institutions, besides the above-mentioned research consequences, has financial policy implications. The changing organizational structure places a different burden on the NFS because of this institutional composition. In West Germany, the financing of nonuniversity research activities is divided between the federation and the states in the proportion 74:26; in the NFS it is 59:41. Also, the average financial provision is around 129,000 DM per staff in the NFS (in West Germany it is 100,000 to 200,000 DM); this formula does not give enough consideration to the investment needs of the NFS if it is to catch up.

For additional information on educational expenditures in West Germany from 1980 to 1991, see Table 8.1.

Instruments for the Reorganization of Institutions of Higher Education and Research

The reorganization committees made a number of suggestions in order to support the restructuring process in the NFS. These measures were proposed because the new states, due to lack of finances and the high burden of rebuilding in other areas, were having difficulties with their restructuring efforts. Additionally, some already apparent unfavorable developments were to be corrected. The proposals have a value of 8 billion DM, an amount unlikely to be forthcoming from public funds.

Renewal Program for Institutions of Higher Education (HEP)

The most important set of support measures in the area of higher education and research are to be found in the "Agreement between the Federation and Federal States about a Common Program of Renewal of Higher Education and Research in the NFS" (HEP), signed on July 11, 1991. This program was expanded with increased financing in July of 1992. The program runs for five years for a total of 2,427 billion DM, provided in annual installments and with the funding divided between the federation and federal government on a 75:25 basis.

The aims are immediate aid for personnel renewal at institutions of higher education; ensuring that qualified scholars remain in the NFS; support of junior scientific personnel; qualification of students and scholars; integration of research into institutions of higher education or into institutions sponsored by the federation and federal states; and increasing investments in higher education in addition to the already existing shared commitment for new building construction at universities.

Specific measures taken to date can be categorized into three groups: personnel renewal; encouraging research potential; and equipment/infrastructure.

1. The personnel measures include financial support for rebuilding departments of law, economics, teacher training (education), and some areas of the humanities and social sciences. It is intended to create "foundation professorships"—that is, professorships appointed directly by the ministry concerned, based on a recommendation from the Higher Education Structural Committee. These are to be paid as in the old federal states; there are a total of about two hundred of these at institutions of higher education and one hundred at specialized colleges. Their location is proposed by the Academic Council. HEP program support is limited to five years. The work of the "foundation professors" is also to be supported by delegated professors and retired professors or those on sabbatical from West German institutions of higher education (110 at universities and 32 at specialized colleges). In addition, another 10 million DM is provided to avoid an exodus of scholars from the NFS. These measures amount to 427 million DM.

Another specific area being supported is that of junior scholars, by provisions for the postdoctoral thesis ("Habilitation," two to three years); postdoctoral support (for research positions in West German institutions of up to two years); the installation of doctoral groups; and for funds for up to 585 gifted persons. These initiatives are expected to amount to around 177 million DM.

In addition, measures have been announced for enhanced training and study to permit undergraduates and graduates to achieve qualifications usable all over the FRG. There is provision for scientific personnel to obtain short-term research scholarships and participate in scientific conferences and symposia. There is also financial support for correspondence courses. Overall, 106 million DM is provided for these tasks.

2.   In order to assist professionally qualified scholars and scholar groups from the former Academy of Sciences, a scholar integration program will be created. Originally, it was intended to employ two thousand scholars in institutions of higher education in the 1992/ 93 academic year. The short duration of this support, together with the prospect of the institutions of higher education having to finance these positions completely (as soon as 1994), led to an extension of the support period to five years (1992–96).

A further measure is to support nonuniversity research. The federal states are to receive support for their financing of Blue List institutions, a directive aimed at relieving the federal states of their half share. The estimate of fifteen hundred positions for scholars in the nonuniversity area has proven to be an underestimate, as the Academic Council recommended forty-five hundred positions. Financial support was increased from 200 million DM to 320 million DM. In addition, 67 million DM has been provided for building repair and equipment renewal. About 990 million DM has been provided to encourage research potential.

3.   A third area supports smaller building projects at institutions of higher education (between 10,000 and 500,000 DM) and the purchase of scientific equipment costing less than the legal petty cash limit for higher-education building support. Funds are provided for acquisition and modernization of publicly owned housing for foundation professors and guest scholars. The fund also helps the introduction of interinstitutional collective library catalogues and their inclusion in West German collective on-line systems, including long-distance checkout. The funds provided amount to 520 million DM.

Supporting Measures The Conference of Ministers of Education of the Federal States suggested measures to alleviate the problem of personnel laid off because of lack of demand. These measures, along with HEP and social and employment support, are expected to mitigate the effects of the personnel reduction. The proposed measures apply to nonprofessorial position-holders such as lecturers and research assistants who have a greater probability of being affected by personnel reduction. The decision to restrict a number of the permanent positions to junior scholars, combined with the planned reduction of positions by 40 percent, is a hard blow for older scholars, as they often have little chance to win an appointment by competition. However, a transitional regulation is proposed for their employment in positions that are to be abolished when they leave the institution of higher education. Therefore, a pool of at least two thousand positions is being created, which permits the employment of academics over fifty years of age in those areas that are expected to grow because of the restructuring process. In addition, they can support the work of persons appointed to new professorships, given the high demands on the latter group's time. The pool is provided for ten years and requires a financial outlay of 825 million DM.

Funds are also to be used to provide additional qualifications and further training, as well as supplementary and advanced training courses. Thus, continued employment is to be made available to those affected by personnel reduction; and the graduates of "old," expiring study courses, especially in vocational and engineering schools, will be given the opportunity of earning a qualification at the specialized college level. This measure will run for three years and is expected to cost over 100 million DM.

Effects of Reunification on Selected Areas

Students In the GDR, students were selected by a rigid and regulated procedure within a governmental framework that established the number of acceptances and offered little choice. From 1960 to 1965, the share of students in the relevant age group grew from 10 percent to 17 percent, but the numbers of entries then stagnated and, by 1978, amounted to little more than 13 percent of the relevant age group. They remained relatively constant at that level until 1989.

With a higher-education degree almost everyone could expect a job, even if the choice was limited and the positions did not necessarily fit their qualifications. Students were organized by study groups and received intensive advising and good social support with basic stipends and dormitory space for all. Dropout rates were low and students graduated on time.

Unification changed the situation; governmental regulation and direction were abolished and social support reduced. Studies can now be planned more independently and individually. At the same time, students criticize the lack of future job security, especially with regard to the practical value of professional skills and the insecurity of future employment possibilities.

Students have also been affected by a "dichotomy of subjects"; on the one hand, sciences and technical subjects have been affected to only a relatively low degree by restructuring while, on the other, the humanities and social sciences have been fundamentally altered. Students of science and technology have to meet new formal criteria, but the content of study has not changed. However, humanities and social-science students face significantly greater changes; they must adapt to new, unfamiliar content. They are concerned about the required academic level, adequate teaching, and a discernible relationship between teaching and job practice.[7]

The students' social situation is characterized by their income and available dormitory spaces. The average income available to a student in the NFS is 630 DM (old federal states, 1,070 DM). The share of students who, because of their parents' low income, receive state scholarships is very high, at 88 percent. The comparatively low income is compensated for by low rents for housing and, in comparison to the old federal states, there is, for the time being, a supply of dormitory space such that 62 percent of the students live there. This share, however, will decrease as a result of expected increases in student numbers; housing could become a big problem in the future.

In comparison to the previous system, both the numbers eligible for higher education and the period between secondary school graduation and the beginning of advanced studies (two to three years) remain the same. However, as secondary education is one year shorter, students are on average one year younger and the duration of studies in the NFS is shorter. These changes, which reduce the duration of studies, coupled with the need for income-generating work, could lead to an increase of problems in the employment situation.

A strong increase in student numbers is expected because of the adjustment of the secondary-school structure and the provision of greater choice at institutions of higher education, consistent with FRG law (Article 12, Constitutional Law). Admissions increased by 20 percent between 1989 and 1990, principally because of the reduction of military service and the matriculation of persons previously excluded from studies. Admissions from the NFS are predicted to grow from 30,900 (1987) to about 72,000 by the year 2000.[8]

Students in the NFS will soon face conditions similar to those in Western Germany; they will enjoy new liberties, but have to face new demands such as planning their own study program, managing their individual time schedule, and developing strategies to cope with new social conditions. There will also be increased student mobility all over Germany in the first part of the 1990s. Western students, now matriculating at Eastern universities, report positive experiences.

Personnel Structure and Employment Higher education and research employment has been characterized by a dramatic reduction in the number of positions, although these differ by area and institution. While nonuniversity research positions are established by the recommendations of the Academic Council, in the area of higher education each individual federal state defines the employment total covered by their financial budgets.

Institutions of Higher Education The institutions of higher education in GDR were characterized by greater numbers employed in student supervision and a quantitatively and qualitatively different group of nonprofessorial personnel. In 1989, there were 131,200 students and 31,745 teachers—a supervisor to student ratio of about 1:5, compared to a ratio in the FRG of 1:17. This difference was mainly due to the high percentage of assistants and lecturers employed in teaching in the GDR. However, responsibilities for counseling, administration, and so forth were fulfilled by other personnel in the FRG.

The number of academic employees (excluding medicine) decreased to 22,432 by the end of 1991, with plans for further reductions to 16,458. The largest cuts involve the nonprofessorial group. However, some challenges to layoffs have succeeded in the labor court; total layoffs may be smaller than expected if this practice grows.

There are difficulties also with the new appointment process. Finances are not yet guaranteed for many planned positions, and many disciplines lack enough qualified applicants, particularly for appointments to professorships. The total number of new professorships was planned to increase by 676 positions over the number at the end of 1991. However, the personnel evaluations were still not completed as of the end of 1992. At Humboldt University in Berlin (December 1992), for example, the screening of professors regarding their involvement in Stasi activities (cooperation with the communist security police) had been completed while other categories of academic personnel still had to be checked. In some humanities departments, West German appointments will be necessary, but a position in the NFS is unattractive to these persons for financial or other reasons. In addition, the current age structure of teachers, both in the old and new federal states, is increasing—20 percent of the professors in the NFS are approaching retirement. At the same time, there is a lack of qualified junior scholars.

Research Institutions The starting point for the restructuring of NFS research and development is the goal of an approximately equal supply of professional personnel in both the GDR and the FRG. The "real" level of East German research and development personnel, using the OECD method of full-time employment units (FEU), was about 140,000 in December of 1989. This was expected to fall to 47,000 FEU by the end of 1992.

Different sectors will reduce employment for different causes. For example, industrial research and development will likely decline from 86,000 FEU to less than 25,000 FEU because of the collapse of old industries (the Academic Council has recommended that nonuniversity-sector employees be reduced from 32,000 FEU to about 11,000 FEU, with 10 percent of the scholars coming from the old federal states). The reduction of positions was smallest through the end of 1992 in the area of higher-education research institutions, although only a small percentage of research was undertaken at these institutions in the GDR. The FEU estimate for institutions of higher education was 14,000 (December 1989); by the end of 1992 the number was expected to be around 10,000 with the possibility of additional cuts because of financial pressures on the NFS. Even assuming that, in 1992, the total number of employees in the NFS was around 5.5 million (versus approximately 9 million in 1990); the proportion of employees working in research and development was only about 60 percent of the comparable number in the old federal states. In industrial research, the comparable proportion was below 50 percent.

The areas of construction and agricultural research were especially affected by the reduction of research and development personnel; the share of research and development positions in these fields was expected to be only 23 percent and 34 percent in 1992. For those disciplines associated with the former Academy of Sciences, reductions were expected to affect about one-half of the staff, with the decrease in the humanities and social sciences, because of their low initial base, being the smallest.

The goal of a balanced regional distribution of research and development potential does not seem to have been achieved. In Thuringia, the shortage of research and development employees was being somewhat corrected, while in Mecklenburg-Prepomerania the real share of research and development employees decreased. These are the weakest regions of the NFS.

Foreign Relations Until 1989, foreign relations for higher education and research, like other areas, were characterized by a high degree of central planning and direction. Foreign relations were limited to exchanges within COMECON, or rather the education of students from socialist developing countries. In 1987 there were 12,745 foreign students. Teachers were hampered by restrictions on travel to the West and, in the humanities and social sciences, by ideological reservations. Attempts to break the international isolation were impaired by ideological obstacles and barriers to the transfer of knowledge on both sides.[9]

Now that the system has been integrated into West Germany, there are two proposed foreign academic goals specifically involving NFS institutions: (1) their integration into international networks as quickly as possible in order to prevent a growing gap in comparison to West German institutions; and (2) continuation of relations with institutions in Southern and Eastern Europe, so that German universities can function as a bridge between East and West:

1. The integration of NFS institutions of higher education into international networks is based on their entry into DAAD (the German Academic Exchange Service, national coordinator and scholarship distributor for different programs). Also, each institution of higher education is establishing academic foreign offices charged with

advising on and carrying out specific foreign activities. The programs covered by the DAAD are now available to NFS institutions of higher education. Long-term project funds are partly supplied through special programs for NFS institutions. In particular, existing partnerships with institutions of higher education in Western countries will continue; support will be provided from a special European Community (EC) fund of ten million ECU for EC programs in the NFS that focus on the integration of the institutions of higher education into the ERASMUS/LINGUA II and COMET II programs; a special program, "Summer Short-term Stipends for English and French," for students and graduates from the NFS, financed by the DAAD, with funds from the Federal Ministry of Education and Science, which, to date, has been a great success;[10] and scholarships for junior researchers in foreign countries within the framework of the HEP (Article 4.2b and c).

2.   Following reunification, the DAAD has taken over the GDR's international programs of higher education in order to carry them through to the end of their agreements. These involve mostly students from Southern and Eastern Europe and developing countries. There were also a number of institutional contacts that have grown over decades and still continue. Only rarely can institutions of higher education uphold their part of the cooperation agreements because of their negative financial situation and staff changes. A financial contribution from the FRG to a partner university must be expected in some cases—for example, cooperation between a German and an East European university, where the latter faces even more difficult financial circumstances.

The future development of NFS foreign relations is overshadowed by the uncertainty implicit in the still ongoing restructuring process. Some of the above programs face other financial problems, such as bureaucratic obstacles, difficulties in acquiring and interpreting information, curriculum changes, institutional restructuring, and personnel insecurity resulting from job cuts, individual financial problems coupled with the problems of institutions and infrastructure, and problems of mobility due to social particularities.[11]

All together, the scope of foreign relations falls short of those found in the old federal states and even those undertaken during the GDR period. Again, it should be noted that there are significant regional differences.

Reflections on the Functions of Institutions of Higher Education

Institutions of higher education are not only part of a system of higher education and academia but also fulfill other specific political, societal, community, and cultural functions. They can be analyzed as contributing to, and forming perceptions of *qualifications*, where the focus is on professional training services that are primarily demanded by the economic system and partly paid for by the financial support of the FRG; *legitimation*, particularly of the political system, which guarantees and regulates the structure and function of institutions of higher education in return for a service of legitimation; *integration*, as individuals receive a status in society by being ascribed a position resulting from their training in institutions of higher education; and *interpretation*, which describes cultural adoption and change based on institutions of higher education, which fulfill this function by generating knowledge through research.[12]

As the economic structure and society undergo rapid change, how far and in what ways are institutions of higher education prepared for transition, and which programs and reforms support that process?

Qualifications The qualification service provided by institutions of higher education must be considered in two parts: First, the transition phase, and how far the training received in NFS institutions of higher education compares with training in the old federal states and is accepted in the labor market; second, the demand for labor in the medium term. Will the proposed structure of higher education meet foreseeable demands? These questions can be answered only with conditional statements because, for the time being, no comprehensive data are available.

Acknowledgment of Degrees The acknowledgment of academic degrees from the old GDR system of higher education is warranted in Article 37.1 of the reunification treaty by general guarantee. An assessment of the value of these different degrees, however, paints a more complex picture. A decision of the KMK (Conference of State Education Ministers) classifies the old degrees of higher education into four types:

1.   Degrees of value (level and profile) equal to that of those received from a university or comparable institution of higher education in the former FRG.
2.   Degrees of value equal to that of those received from a university or comparable institution of higher education in West Germany, but where the training was aimed at the economic and social system of the GDR, so that significant system-related differences exist in regard to study content.
3.   Degrees of value (level and profile) equal to that of a degree received at a specialized college.
4.   Degrees of a level equal to a degree from a specialized college but where the training was aimed at the economic and social system of the GDR, so that significant system-related differences exist in regard to study content.

This categorization was applied to 330 different subject areas and evaluated by comparing the study plans and examination rules with those of similar courses in Western Germany. Only fifty percent of graduate degrees were judged to be equal in level, and they will therefore be disadvantaged in competing in the labor market. The differences in "profile" can, however, be decreased by carefully directed further qualification and training.

The first analyses of the labor market in the NFS demonstrate that the rate of unemployment is significantly lower for persons with an advanced degree than for persons with other types of training. However, employees' perceptions in this regard are changing. In November 1990, 80 percent of employees with an academic degree believed that there was a correspondence in qualifications and skills; one year later it was only 71 percent. Aggregated labor-market data for the NFS does not permit an examination of specific fields.

Prognoses of Need for Qualification If the number of gainfully employed persons is estimated to be 6.8 million in the year 2000 in the NFS (using the IAB-Westphal-Modell; see Fuch et al., 1991),[13] then, relying on past ratios, 380,000 students (including 70,000

beginning students) corresponds to the level found in the old federal states in 1990. To train this number of students, about 33,000
full-time academic positions might be required. This number is similar to calculations based on admission estimates in the NFS (Weegen
1992). However, the proposed numbers found in individual federal
state plans are much lower. Scientific personnel are to be reduced to
16,500 by 1993, and it is not expected, for budgetary reasons, that
the number of positions will be doubled by the year 2000. So it is
likely that neither the demand for qualification nor admissions will
satisfy expected demand. At present a regional approach is being
used which assumes that the individual federal states will provide
study places for the students of their own federal state. Exchanges
between new and old federal states are not considered, nor are developments in the EC.

However, even these recommendations of the Academic Council and the federal states are not being fulfilled. The Academic Council
commented in a declaration of July 3, 1992:

*The necessary structural and personnel renewal of the institutions of
higher education is currently suffering mostly because the liquidation of
obsolete structures has not been adequately combined with a quality-
oriented reconstruction. The implementation of the recommendations of
the state structural Committees therefore is proceeding only slowly. In
addition they are impaired, in some cases, by inappropriate decisions
and unsatisfying compromises.[14]*

Legitimation The higher-education institutions' legitimation function is of great importance for the process of Germany's inner reunification. This concerns, in particular, the demands and degrees of
freedom that the system of higher education in the NFS ascribes to
individuals and social groups within the system. The attitudes toward the GDR government were characterized by adaption and
impediments to criticism and creativity. The heterogeneous democratic concept, designed during the first months of 1990, are now
encountering FRG legislation. Together with institutional autonomy,
the legislation provides for participation of certain groups in the
self-governance of institutions of higher education. However, NFS

universities face a complicated set of rules and cases which they do not understand because they lack both the particular history and context. This is not favorable for active participation. Moreover, as the key positions in ministries and universities are filled with Western experts, the remaining Eastern personnel participate less in decision-making. Among some students and staff, the experience of restructuring, especially the liquidation of institutions, has created an attitude of incomprehension and helplessness vis-à-vis decisions of the state. This mood impedes active participation to the point that some talk of the process as "Westernization" or "colonization."

The withdrawal from political structures that can be found in society in general is apparent in a number of academic institutions. An example is the development of the "Conference of Student Unions" in the NFS. The conference, founded in 1989 as an interest group of graduate students, tried to influence political decisions at many levels. By 1992 the conference was restructured to form a loose information network, indicating that the attempt to establish a critical potential within universities could only partially be realized.

Integration Status distribution is being allocated in two ways; the two forms of social differentiation are social background and gender-related inequalities.

Social Background In both parts of Germany, academicians made up 8 percent of the total population. However, self-recruitment of academicians was more strongly pronounced in the GDR than in the FRG. In the NFS, 54 percent of new students in 1990/91 had at least one parent with an advanced degree. Between 1979 and 1989 the proportion of students who had at least one parent with an advanced degree increased from 28 percent to 52 percent, while at the same time the share of students from skilled workers' families decreased from 24 percent to 14 percent. In the FRG, the share of students from academic families was reduced from 36 percent in 1960 to 25 percent in the late 1980s, and the share of working-class children increased from 5 percent to 15 percent. The number of students admitted in 1990/91 rose to 31 percent. This increase can be attributed to more restrictive regulations with respect to individual financial support in the FRG.

Any data comparison must take into account that access to higher education was strongly regulated early on in the GDR and that higher education, unlike in the FRG, did not pay off socially and financially. Future trends should not be estimated, as admissions in the NFS continue to be based on selections from the old GDR school system.

Gender-Related Inequalities Since the mid 70s, women in the GDR, as a proportion of all students, made up slightly more than 50 percent, which corresponded to their population share. In the FRG, the share of women increased to 38 percent for the first time in 1982 and has remained constant since.

Beyond the first degree, the share of women earning further academic qualifications decreased step by step in both the GDR and FRG. Only 5 percent of lecturers and professors are women. Both systems therefore demonstrate structurally caused disadvantages to women in an academic career; in both states, female students are strongly underrepresented in the sciences and technical subjects; they represent a higher percentage than male students in social, cultural, and educational subjects.

Initial information, following reunification, indicates that the higher share of women in the NFS will quickly adjust to the lower proportion found in the old federal states. Female admissions in 1990 were only 38.5 percent, leading to a decrease in the number of female students in the NFS to 45.2 percent. Some part of the reduction is unlikely to be repeated and can be attributed to a reduction of military service for men. However, surveys among secondary-school graduates with higher-education admission qualifications show that women's intention to study has decreased by 9 percent to 69 percent, while men's has remained constant at 79 percent (1990/91).[15]

Although gender equality is a legal requirement, women are scarcely represented in the committees in charge of the restructuring of institutions of higher education or among new appointments. The HEP and its regulations have failed to strengthen the position of female scholars in the NFS, as the HEP promotes the areas of academia that are male domains and where women are rarely found in the old federal states.[16]

Interpretation The structure of GDR institutions of higher education was characterized by a single-minded direction toward professional training and an implicit reduction of research capacity by the transfer of research areas into research institutions outside of universities. The research that remained at universities often had to follow central research plans. The current process expects to improve this previously unbalanced relationship in terms of teaching research and staffing. These plans, however, coincide with a dramatic cut in the number of positions at institutions of higher education, as described above. In this uncertain situation, it can be observed that many highly qualified junior scholars and researchers leave institutions of higher education when offered another position; they can expect better job security and higher salaries in the nongovernmental sector. This exodus is a major problem for institutions of higher education; the state-funded integration programs do not provide enough incentive to redress the problem.

The Academic Council has assumed that the first priority is the improvement of qualifications, so the rebuilding of teaching programs was the first focus. At the same time, however, they hope that a foundation is being laid for future research at institutions of higher education by means of changes to the staffing structure, by altering permanent positions (those of lecturers and assistants) to nontenured positions, and by developing programs for doctoral candidates and participating in collective research projects with similar institutions in the old federal states.

An important structural adjustment involves arrangements for closer cooperation between nonuniversity research institutions and institutions of higher education. One way to accomplish this is by appointing leading scholars to both types of institutions, where they are expected to teach as well as do research. This will allow institutions of higher education, which complain about insufficient research equipment and access to laboratories, as well as doctoral and diploma candidates, to obtain access to nonuniversity research results.

Back to the Mainstream—A View from 1995

Looking back on the transition period of the early 1990s from mid-1995 shows the astonishing amount of change and reconstruction that has occurred. Most visible is the change in buildings and equipment. Historical university buildings have been carefully restored, decayed rooms renovated, new functional furniture brought in, library stocks drastically enlarged, photocopy machines made available everywhere, laboratories and workshops equipped with the most recent technology, and computer networks introduced; researchers in the eastern part of Germany now navigate in the internet like their counterparts in the West.

Research institutions and universities have been integrated into the nationwide model that is shaped according to the former West German patterns. There is now only one German system of higher education and research institutions. When there are regional particularities in the eastern institutions, they follow more the traditional regional identity of the state (like Saxony or Thuringia) in which the institution is located than a common "East German" pattern.

The restructuring process concerning personnel has resulted in a big West-East transfer for some disciplines, for example philosophy, education, history, economics, and law, at least for professorial positions. On the other hand there is an East-West transfer, much smaller, of top level researchers in some other disciplines like medicine. Those academic people in the five new states who lost their positions during the restructuring period perceive themselves as losers in the unification process and express their views, sometimes vividly, in private correspondence and public interviews. Students tend to choose their college more and more according to subject specificity or regional preferences than according to East-West categories.

Being back to the mainstream has a lot of positive aspects. However, it means that the institutions in the eastern states of Germany also share the problems and deficiencies of the university and research system in the western part of the country. The transition period meant a huge restructuring and improvement effort, but it was not a reform movement in the sense of implementing innovative ideas. A prominent counter-example is the newly founded "Europe University Viadrina" at Frankfurt/Oder that established in some disciplines an innovative, internationally, and interculturally oriented curriculum, in part stimulated by some of the ideas that the president of the institution, Hans N. Weiler, brought with him from Stanford University.

Notes

1.   C. Melis, "Zur Rolle der Akademie der Wissenschaften im Wissenschaftls-

system der DDR, Gewerkschaft Erziehung und Wissenschaft (GEW), "
Materialien und Dokumente Hochschule und Forschung, 65, S.72–87, 1990.

2. "Wissenschaftsrat, Empfehlungen zu Hochschulstrukturkommissionen und Berufungspolitik an den Hochschulen in den neuen Ländern und in Berlin," 16 (11), 1990.

3. "Wissenschaftsrat, Empfehlungen zu den Ingenieurwissenschaften an den Universitäten und Technischen Hochschulen der neuen Länder," 5 (7), 1991, Drs. 325/91, p. 11.

4. Ländergesetze:
    • Sächsisches Hochschulerneuerungsgesetz vom 25.7.1991.
    • Sächsisches Hochschulstrukturgesetz vom 10.4.1991.
    • Brandenburgisches Hochschulgesetz vom 24.6.1991.
    • Entwurf zum zweiten Hochschulstrukturgesetz des Landes Sachsen-Anhalt.

5. W. Pohl, "Kürzung ohne Konzption, Krise der Forschungsfinazierung und das Modell FhG." *Forum Wissenschaft*, 4/91, S. 50-55.

6. "Wissenschaftsrat, Empfehlungen zum Aufbau der Wirtschafts und Sozialwissenschaften an den Universitäten/Technischen Hochschulen in den neuen Bundesländern und im Ostteil von Berlin," 17 (5), 1991, Drs. 254/91, S. 36.

7. U. Heublin and F. Kamzemzadeh, "Studieren in den neuen Ländern—eine Untersuchung der Studienbefindlichkeit unter strukturell veränderten Bedingungen" (Hannover: Hochschulinformationssystem (HIS) Kurzinformation A10/91, 1991).

8. M. Weegen, "Eckdaten für eine aufgabengerechte Personalausstattung der Hochschulen in den neuen Bundesländern" (Frankfurt: 1992), p. 12.

9. B. Last and H.-D. Schaefer, "Die internationale Dimension der Hochschullandschaft: Ausländer- und Auslandsstudium unter besonderer Berücksichtigung der Wissenschaftsbeziehungen zu Osteuropa, Projektberichte 4/1991" (Berlin: Projektgruppe Hochsculforschung, 1991).

10. Deutscher Akademischer Austauschdienst (DAAD), Jahresbericht 1991 (Bonn: 1992a).

11. S. Manning, "EG-Bildungsprogramme—ein Anreiz für die neuen deutschen Bundesländer" (Berlin: unpublished manuscript, 1992).

12. V. Lenhart, "Herausforderungen an die Hochschulen in der Bundesrepublik Deutschland angesichts der deutschen Einigung und des europäischen Binnenmarkts," MS Heidelberg, 12 (4), 1991.

13. J. Fuchs, u.a., "Erste Überlegungen zur künftigen Entwicklung des Erwerbspersonenpotenials im Gebiet der neuen Bundesländer," MittAB 4/91, s., pp. 689–705.

14. Wissenschaftsrat, "Empfehlungen zu den Geisteswissenschaften an den Universitäten in den neuen Ländern," 3.7.1992 Drs. 812/92.

15. F. Durrer, "Ausbildungswahl der Abiturienten 91 aus den neuen Ländern" (Hannover: Hochschulinformationssystem (HIS) Kurzinformation A9/91, 1992).

16. A. Burkhardt and G. Stein, "Gleichstellungspolitik an den Hochschulen— Gesetzliche Grundlagen und Realität." Projektgruppe Hochschulforschung in Berlin (Berlin: Karlshorst, MS, 1992).

Part Three

Future Directions

Chapter Nine
World Bank Lending for Higher Education and Research
Lessons and Implications for Eastern Europe

Thomas Owen Eisemon

The World Bank's lending strategy for human resource development in Eastern Europe emphasizes the importance of restructuring education and training systems to facilitate the transition to market economies. Projects have generally combined initiatives directed to transforming the organization and content of vocational and technical training, with support for efforts to break down the compartmentalization of the systems of higher education and research.

While projects reflect the diverse circumstances and different priorities of the borrowing countries, there are many commonalities. Bank investments support the "structural adjustment of the human capital producing sectors . . . to the new political and economic realities," particularly "the change in the function of labor markets from social welfare and redistribution to economic productivity and efficiency."[1] The introduction of market mechanisms has led to sharp reductions in industrial output throughout the region and has combined with decreased subsidization and privatization of some public enterprises, increasing unemployment. This has focused concerns on the need for comprehensive national strategies for retraining the labor force and restructuring the relationship between the education and the employment sector, as well as for maintaining and eventually increasing human resource and capital investment in modernizing the technological bases of production systems.

This chapter examines bank strategies for reforming and rehabilitating systems of higher education and research in Eastern Europe in the context of its worldwide lending experience in higher education since 1963. The overall objective is to show how the bank is attempting to apply lessons learned in other geographic areas to its evolving lending program in Eastern Europe. The first section below surveys the kinds and distribution of higher-education investments by the bank from 1963 to 1991. The second considers projects in Brazil, China, and Hungary that have significantly influenced the way the bank has understood the key issues involved in restructuring systems of higher education and research in Eastern Europe. The final section draws attention to special circumstances that the bank must take account of if its present and future lending activities are to be effective.

Patterns of Investment

Since 1963, the World Bank has had a prominent role in supporting the development of systems of higher education in Latin American, Caribbean, African, Middle Eastern, Asian, and many European countries.[2] In all, 281 higher-education projects with 432 higher-education components have been supported, amounting to a total bank investment of U.S. $5,075,900. That represents almost one third (37 percent) of lending for education during the period 1963 to 1991. World Bank investment in higher education has grown from an average of 17 percent of lending for education in the 1960s to 37 percent since 1986.

Between 1963 and 1975, a majority (61 percent) of projects supported universities. In the early 1970s, the justification for large donor and government investments in higher education, particularly in university development, were being questioned. Rate of return analyses suggested that developing countries were "overinvesting" in higher education and that resources should be redirected to primary education. This theme was amplified in World Bank education research and policy statements throughout the 1970s and early 1980s. The 1971 Education Sector Policy Paper proposed more emphasis on primary and even nonformal education.[3] The 1974 Education Sector Working Paper criticized the disproportionate allocation of education resources to secondary and higher education that served the modern sector,[4] resulting in underfinancing of basic education, which was both more efficient and more equitable.

The 1980 Education Sector Policy Paper focused on equity issues and on expanding access to basic education within the framework of measures to promote cost-effectiveness and external efficiency.[5] It raised concerns about reliance on manpower forecasting and the enthusiasm of many developing countries for vocational training, though it favored investments in polytechnics and other forms of technical training as an attractive alternative to high-cost university studies. The number of World Bank investments involving universities declined between the mid 1970s and mid 1980s. Nevertheless, the volume of lending grew more than 50 percent, and higher education's share of education lending increased from 30 percent to 43 percent.

The volume of lending for higher education has continued to grow, although its share of education investments has declined somewhat. Nearly all of this relative decline can be attributed to rising support for primary education. In other words, the competing requirements of the higher-education and primary-education subsectors have been accommodated by the expansion of assistance for education generally.

Investment in European and Middle Eastern and especially East Asian countries has grown more rapidly than in African or Latin American and Caribbean countries. In the period 1986–1991, East Asian countries—principally China and Indonesia—accounted for 33 percent of project investments and nearly half (47 percent) of lending for higher education. Lending to South Asian, European, and Middle Eastern countries has also grown in recent years.

World Bank lending for higher education has been mainly directed toward institutions that supply educational systems with teachers to facilitate expansion of school enrollments or the productive sectors with technicians. Since 1986, for instance, about a third (30 percent) of investments in higher education have supported teacher training and 24 percent have been directed to polytechnics or technical-training institutions. The proportion of university projects has increased slightly since the early 1980s to 34 percent in the period 1986–91. Support for national scientific institutions that carry out advanced training and research is very recent. Together with universities, which in many developing countries employ most scientists and engineers engaged in research and development, World Bank lending to these institutions between 1986 and 1991 accounted for nearly half (46 percent) of project investments and 58 percent of lending for higher education (see Tables 9.1 and 9.2).

Table 9.1 Total Lending for Higher Education by Region, 1963–91 (millions US$)

Region	1963 $	1970 Percentage	1971 $	1975 Percentage	1976 $	1980 Percentage	1981 $	1985 Percentage	1986 $	1991 Percentage	Total $	Percentage
Africa	15.6	30.2	43.3	13.4	69.7	12.0	111.7	7.3	279.0	10.8	519.3	10.2
East Asia	11.9	23.0	108.0	33.0	214.4	37.0	1,103.1	72.0	1,219.0	47.0	2,656.4	52.0
South Asia	17.4	34.0	54.0	17.0	22.6	4.0	67.9	4.0	569.3	22.0	731.2	14.0
EMENA	3.0	6.0	104.9	33.0	264.5	46.0	149.9	10.0	357.4	14.0	879.7	17.0
LAC	3.7	7.0	12.4	4.0	7.4	1.0	100.4	7.0	165.5	6.0	289.4	6.0
Total	51.7		322.6		578.6		1,532.9		2,590.1		5,075.9	

Table 9.2 Total Lending by Type of Institution Supported, 1963/91

Institution	1963/70			1971/75			1976/80			1981/85			1986/91			Total		
	No.	(millions US$)	Percentage	No.	(millions US$)	Percentage	No.	(millions US$)	Percentage	No.	(millions US$)	Percentage	No.	(millions US$)	Percentage	No.	(millions US$)	Percentage
Universities	10	$32.3	63%	25	$187.6	58%	26	$268.0	46%	23	$824.5	54%	38	$899.7	35%	122	$2,212.1	44%
Science and Technology research institutes	0	$0.0	0%	1	$2.8	1%	0	$0.0	0%	3	$129.5	8%	13	$604.8	23%	17	737.0	15%
Polytechnics	3	$4.1	8%	9	$32.6	10%	6	$66.9	12%	9	$149.5	10%	10	$566.5	22%	37	$819.6	16%
Technical institutes	7	$8.1	16%	30	$75.2	23%	34	$189.4	33%	19	$240.1	16%	17	$209.9	8%	107	$723.1	14%
Teacher-training institutions	15	$7.2	14%	33	$24.4	8%	34	$53.7	9%	34	$189.4	12%	33	$309.2	12%	149	$548.0	12%
Total	35	$51.7	100%	98	$322.6	100%	100	$578.6	100%	88	$1,533.0	100%	111	$2,590.1	100%	432	$5,075.9	100%

Lending Experiences, Lessons Learned

Rehabilitation of China's System of Higher Education and Research
The bank's lending program to China illustrates the range of activities it supports, and the successes of a comprehensive, integrated approach to rehabilitation of higher education systems, as well as some of the difficulties experienced in transforming the training and research systems of countries with socialist economies. Nine projects have supported strengthening the different tiers of the higher-education subsector within the framework of China's Four Modernizations Plan of 1980. The first project (1980) provided $285 million for improving scientific and technological training at about a third of the country's elite national universities.

World Bank funding facilitated construction or rehabilitation of university laboratories and libraries, updating instructional and research programs with expert foreign scientific assistance, and upgrading the professional qualifications of academic staff through foreign training. Project implementation was guided by a Chinese Review Commission composed of many of the country's leading scientists as well as by an International Advisory Panel. Somewhat similar international advisory structures were developed and implemented successfully in earlier Korean science and engineering projects.

Two agricultural training projects (1983, 1984) and a polytechnic/television university project (1983) were concerned with expansion of agricultural and technical universities and specialized training and research institutions like the National Rice Research Institute. A second university development project was approved in 1985 to strengthen economics and engineering programs at thirty-five national universities supervised by government ministries. But neither this project nor the previous ones were intended to address in any significant way the increased pressure for expansion of the country's very selective system of higher education. They were designed mainly to enable China's elite institutions of higher education to catch up in scientific training and research after a long period of professional isolation during the Cultural Revolution.

The Provincial Universities Project (1986) embraced the largest, most diverse component of the higher-education subsector—60 of the country's more than seven hundred second-tier medical, technical, agricultural, and comprehensive universities. Although the project gave importance to quality-improvement measures such as staff development, curriculum reform, and investments in laboratories and libraries, more importance was placed on expansion of enrollment than in early projects. The project was implemented in the context of a radical decentralization of the State Education Commission's responsibility for financing and directing the growth of higher education at the provincial level.

Subsequent projects have supported university presses to increase the availability of instructional texts (1989) and one of the most recent, the Key Studies Development Project (1991), attempts to strengthen postgraduate training and research at 133 State Key Laboratories and Special Laboratories either at selected national universities or in institutes administered by the Chinese Academy of Science.

Over the course of more than a decade and through multiple project investments, China's system of higher education has been strengthened considerably. At the end of the Cultural Revolution, the government embarked on a massive effort to join the newly industrialized East Asian countries in terms of the production of scientists and engineers, and to promote technological innovation in agriculture and industry through increased access to scientific and technological expertise from other countries. World Bank assistance has facilitated China's reentry into the international mainstream in many fields of scientific training and research. By the mid-1980s, the mainstream scientific output of Chinese scientists and engineers had grown considerably, reflecting the country's new scientific, educational, and developmental priorities and increased international scientific cooperation. The qualitative expansion of scientific and technological training has been impressive, so much so that since 1988 the State Education Commission has frozen university intake out of concern for the qualitative implications of rapid expansion of enrollment.

There were also important curriculum reforms in engineering, education, and applied social sciences. For instance, in engineering, the number of different degree programs has been reduced from 665 to 100. Programs are still overly specialized in comparison to those of many industrialized countries, however. Despite Chinese efforts to delink engineering education from manpower planning

and create a labor market for engineers, state-owned enterprises remain the principal employer of engineering graduates, and the configuration of an engineer's professional responsibilities changed slowly
until very recently.

The reform of programs in economics and financing has been,
perhaps, more successful. These were radically changed with profound long-term implications for the training of policy cadres that
have played a central role in transforming the country's economy. In
1991, undergraduate programs in economics and finance at 250
universities and colleges were directed by the State Education Commission to begin using a new market economics syllabus developed
by expert panels of Chinese and foreign scholars. "This would not
have been possible," Hayhoe notes, "without Chinese consensus on
both the need and direction of curricular reform, including recognition that western social science training models could be adapted to
the Chinese situation."[6]

Nevertheless, the World Bank has largely been unable to enter
into a policy dialogue with the government on financing and efficiency issues that are central to the sustainability of its large investments. Ongoing negotiations with the government over reducing the
large number of provincial and municipal universities and colleges
with small enrollments and many staff has so far not led to major
policy reforms, although the 1992 Education Development in Poor
Provinces Project supports experimental pilot programs to increase
internal efficiency by encouraging mergers of institutions, reorganization of programs, and the reassignment of redundant teaching staff.

Supporting Scientific Training and Research in Brazil Like many
Eastern European countries, Brazil is a major producer of science
and technology, scientists and technologists. But it has not derived
full benefit from its substantial research and training capacities for
several reasons. The country's research and development capabilities
are mainly concentrated in the public sector.[7] There is a low level of
private investment in research and development and, because of large,
protected, oligopolistic internal markets, few incentives for technological innovation by private firms or state-owned enterprises.[8] Moreover, the country's workforce is, overall, poorly educated and largely
unskilled.

For many years, the World Bank provided little assistance to higher education apart from what was necessary to support Brazil's agricultural research system and expand research on nonfossil fuels and the ecology of the Amazon basin. No rationale was developed for greater involvement in the system of higher education. Indeed, from a sectoral perspective, there was little justification for major investment.

The Science and Technology Project approved in 1985 reflected a radical change in the World Bank's approach. Better exploitation of the country's agricultural, mineral, and other natural resources were seen as requiring increased investment in the country's infrastructure for scientific and technological training and research. A rationale for investment was developed, focusing on the implications of investments in research and training for economic growth and the country's poor performance on science indicators relative to OECD (Organization for Economic Cooperation and Development) and many Asian countries.

Sector work identified the principal weaknesses of Brazil's systems of higher education and national research: (a) dispersion of scarce resources "across too many discrete activities" with the result that projects are inadequately funded; (b) "mismatch between availability and orientation of scientific and technological expertise and the unexploited natural resource base"; (c) poor scientific support services and access to scientific information; and (d) ineffective mechanisms for setting scientific and educational priorities and funding projects.[9]

A human-resource subprogram to expand postgraduate and diploma training in institutions of higher education located in regions rich in natural resources but scientifically and educationally less developed was funded, together with a large targeted-research-grant program in applied scientific fields open to researchers in universities and governmental scientific institutions. Procedures for peer review were approved in 1990 to "consolidate the still fragile institutional reforms made in the first project, such as open competition, peer review and decentralized planning."[10] The project continues the focus on the public-sector research system and "activities which the private sector will not finance." Additional funding is provided to the fields in the applied sciences supported by the earlier project, and two more have been added, materials science and environmental studies. Almost half of project funding will be expended for the purchase of scientific equipment.

Much of the rationale presented for these project investments rests on the assumption that Brazil is not producing enough high-quality science or well-trained scientists and engineers. The problem is insufficient research and training capacity rather than insufficient utilization of the country's public higher-educational and research assets by productive sectors. Yet the appraisal report for the most recent project draws attention to many areas of concern. The public sector, it notes, may finance as much as 90 percent of total science and technology investment, a much higher proportion than in many OECD and East Asian countries.[11] While state funding has declined from the early 1980s as a result of continuing austerity, federal-government spending on basic research has remained stable, reflecting "the political power of the Science and Technology community."[12] Ten percent of national investment in advanced scientific training and 40 percent of competitive research funding, most of which is captured by universities, is derived from World Bank loans.[13]

Macroeconomic reforms introduced by the federal government in 1990 have profound implications for whether the country will ultimately benefit from investments in expanding its systems of higher education and research. Protected internal markets are being opened and firms encouraged to compete, with tariffs being reduced substantially. State-owned enterprises, which account for a substantial share of national research and development investment, will soon be privatized. Such measures might lead to greater utilization of the research and training capabilities of universities and governmental scientific institutions. Much will depend on the policies the country develops to stimulate private investment in research and development and also investment to increase the educational and skill level of the labor force.

Reforming the System of Higher Education and Research in Hungary The political and economic context of the bank's 1991 investment in reform of Hungary's system of higher education and research is more dynamic than in either China or Brazil. A radical reform of the structure and financing of higher education and research is being supported in order to expand access to higher education, encourage curriculum reforms, and unify the country's systems of higher education and research, which are presently the responsibility of different ministries and scientific academies.

In conjunction with the implementation of the World Bank project, a Higher Education Law was presented to the national parliament in 1992 to reorganize the public higher-education system, allow public institutions to levy fees, change the procedures for allocating state support to students and institutions, legitimate the establishment of private institutions, and create funds to support research, institutional rehabilitation, and innovation, as well as to establish new policy structures to guide the growth of the system of higher education.[14]

The Higher Education Law brings all public and private higher education within the authority of the ministry responsible for higher education, which will be advised by a Committee for Higher Education and Research composed of representatives of the various ministries concerned with higher education, university rectors and administrators of governmental scientific institutions, and local and foreign experts. The committee will formulate norms for financing public and private institutions and make recommendations to parliament on public expenditure for higher education through the ministry.

Although higher education will continue to be supplied primarily by the state, public and private institutions will compete for state support. Support will be distributed to institutions and students through various funds. Students will receive payments from the state Student Fund for part of the costs of accommodation, boarding, textbooks and, in addition, loans to pay these and other costs, interest free, for ten years. A Tuition Fund will provide support to institutions based on their efficiency and other parameters of performance, as well as the costs of the programs they offer. Budgetary incentives will foster qualitative improvements and curriculum innovations.

The Higher Education Research Fund will selectively support proposals from institutions that fall outside the mandate of other national research councils, while the Facilities Fund can be accessed by institutions for extraordinary capital investments. Universities and colleges will be allowed to determine and allocate intake except in the case of certain professional faculties (such as medicine, dentistry, veterinary medicine, and so forth), whose enrollment will be controlled by the Committee for Higher Education and Research. The institutions will be able to obtain additional income from private sources (without reduction of their operating budgets), set salaries and wages of academic and nonacademic staff, and manipulate other aspects of their cost structure subject to minimum accrediting standards adopted by the Committee for Higher Education and Research to determine eligibility for support from the Tuition Fund.

The project shares some features of the Chinese and Brazilian projects described above, especially the reliance on open competitive-funding mechanisms to motivate reforms, with the difference that the scope of the reforms supported is much broader. A Catching Up with Higher Education in Europe Fund will be established with a World Bank credit to provide incentives to institutions to introduce new programs, integrate the research functions of governmental scientific institutions with the teaching activities of universities, and promote collaboration with other European institutions. The fund provides interim support for the comprehensive reforms envisaged in the Higher Education Law.

Implications for Future Lending in Eastern Europe

While it is premature to speculate about the outcome of the bank's project in Hungary, there are several lessons that can be drawn from the earlier investments in China and Brazil in fostering reform of higher education and research that are pertinent to other countries whose economies are in transition.

First, though a great deal of change can be accomplished by creating incentives and providing funding for institutional reforms, the macroeconomic environment must be supportive. That is particularly the case for reforms designed to increase the curricular flexibility and responsiveness of training programs to emerging labor-market needs. Where progress toward creating open markets and expanding the role of the private sector is slow, curriculum reforms that anticipate changes in the labor market will be difficult to carry out, as the bank's experience with engineering education in China illustrates.

In scientific and technological fields, the problem is compounded by the sharp decline in industrial output that usually accompanies market reforms. At least initially, open markets do not generate new investment in research and development activities or much demand for scientists and engineers, whatever their training. Unemployment of scientists and engineers has risen rapidly throughout Eastern Europe, and enrollments in scientific and technological fields have fallen in several countries. This suggests that reform efforts should concentrate on developing programs in fields in the applied social and natural sciences, especially economics, management, and the environmental sciences, that were largely neglected during the socialist period and for which there is already evidence of a growing demand for graduates.

Second, the establishment of unitary, transparent, competitive-funding mechanisms to overcome the fragmentation of the systems of higher education and research characteristic of many Eastern European countries is not likely to produce a radical change in the structure of these systems in the absence of more fundamental financing reforms of the kind contemplated in Hungary. In China and Brazil, funding mechanisms for competitive research and training have the more limited and more achievable objective of increasing international scientific cooperation and promoting institutional collaboration. They have been effective in sustaining research and training activities in scientific and technological fields under conditions of resource scarcity. However, they are clearly not a long-term solution to the compartmentalization of scientific training and research. A change in the way core budgets for teaching and research institutions are generated and allocated is needed.

The strategy in Hungary is to diversify sources of support for training and research and to supplant existing funding mechanisms with new funds administered by agencies that will enjoy considerable autonomy vis-à-vis the ministries responsible for universities and governmental scientific institutions. Predictably, the proposal has provoked strong opposition from the sectoral ministries, such as agriculture and health, which operate their own training and research institutions, as well as from the scientific academies.

Third, prospects for reform are not likely to improve without greater political stability and legitimacy and more attentiveness on the part of governments and the bank to the democratic context of the reform process. The scientific communities of Eastern Europe are politically powerful, as are those in China and Brazil, but their claims on scarce public resources are generally weaker, largely because of their formerly privileged status during the socialist period. Nevertheless, reforms are no longer promulgated; they must be politically negotiated with the parties concerned.

The establishment of participatory, parliamentary forms of government after 1989 led to greater political autonomy for higher education and scientific institutions. The governance and management of these institutions was democratized and state intrusion reduced considerably except, significantly, insofar as their financing is concerned. Meaningful autonomy in how institutions raise and utilize resources and in regard to matters affecting their costs—such as levels of remuneration—is essential to both financing and curriculum reforms. But mechanisms of accountability acceptable to higher education and scientific institutions as well as to the state will be necessary to support further devolution of governmental control.

An important legacy of the socialist period is a pervasive (and well-founded) distrust of the executive organs of government, especially by elected parliamentarians. This is exacerbated both by the proliferation of political parties and interest groups and by frequent changes in government. The result is that legislation involving reform of higher education and scientific institutions is highly detailed, crafted to circumscribe governmental authority in implementation; it receives intense political and public scrutiny. This has led to delays in passage of important reform legislation in a number of Eastern European countries.

Finally, even under the most favorable circumstances, multiple investments will be required to support the reform process in Eastern Europe. The World Bank will have to work with governments in formulating a long-term comprehensive investment strategy, as it did in China throughout the 1980s, supported by a series of project investments directed simultaneously to different institutional components of the higher-education and research system. Funding for curriculum reform, rehabilitation of instructional and research infrastructure, staff training, and other qualitative improvements must be mutually reinforcing and provided in conjunction with increased national investment, both public and private. High priority should be given to assisting governments to strengthen the management of their higher-education and research systems and institutions.

Notes

1. R. Harbison, "Human Resources and the Transition in Central and Eastern Europe" (Washington, D.C.: World Bank, 1991), pp. 7–8.
2. T.O. Eisemon, "Lending for Higher Education: Analysis of World Bank Investment 1963–1991," PHREE Background Paper 66R (Washington, D.C.: World Bank, 1992).
3. "World Bank, Education Sector Working Paper" (Washington, D.C.: World Bank, 1971).
4. "World Bank, Education Sector Working Paper" (Washington, D.C.: World Bank, 1974).
5. "World Bank, Education Sector Policy Paper" (Washington, D.C.: World Bank, 1980).
6. R. Hayhoe, *Chinese Universities and the Open Door* (London: M.E. Sharpe, 1989), p. 62.
7. S. Schwartzman, *Space for Science: The Development of a Scientific Community in Brazil* (University Park: Pennsylvania State University Press, 1991).
8. C.J. Dahlman, "Foreign Technology and Indigenous Technological Capability in Brazil" in M. Fransman and K. King, eds., *Technological Capability in the Third World* (London: Macmillan, 1984); Schwartzman, *op. cit.*
9. World Bank, *Brazil: Project for Science and Technology (Staff Appraisal Report)* (Washington, D.C.: World Bank, 1985).
10. World Bank, *Brazil: Science Research and Training Project (Staff Appraisal Report)* (Washington, D.C.: World Bank, 1990), p. 12.
11. *Ibid.*, p. 34.
12. *Ibid.*, p. 35.
13. L. Wolff, "Higher Education Reform in Brazil" (Washington, D.C.: World Bank, 1991), mimeo.
14. Coordination Office for Higher Education, "Concept for Higher Education Development in Hungary" (Budapest: COHE, 1991), mimeo.

Chapter Ten
The Way Ahead
Lessons and Prospects for the Future

Barry Lesser and A.D. Tillett

The papers in this volume have described the national experiences of several states of Central and Eastern Europe (CEE) and the Former Soviet Union (FSU) in dealing with the science and technology sector, and higher education more generally, in the wake of the end of the communist era and the dissolution of the USSR. Many of the papers do not describe developments beyond 1992, a result of when the papers were first written. This is not a serious weakness, however. What is most important in all of the papers is the analysis of the problems facing the systems of higher education at the outset and the description of the difficulties that have confronted efforts to introduce change and, indeed, to define new directions. Dealing with the initial sets of problems would be a long-term proposition under the best of circumstances. When the obstacles to change are factored into the equation, there is little reason to expect that the events of one or two subsequent years will dramatically change any of the major conclusions presented in the papers here. Progress is being made, but only slowly. For the countries examined in this volume, the basic challenges that have been identified still remain largely to be met. In this final chapter, an effort is made to summarize these basic challenges.

Results

A first and obvious conclusion that emerges from a reading of the papers is that the system is under great stress from both within and without. The internal pressures largely emanate from the legacy of the Soviet system of science and technology and higher education; they encompass such problems as inappropriate fields of specialization, imbalances in research versus teaching fields, poorly qualified professional staff, excess personnel, outdated equipment and technology, space constraints, lack of research materials (especially library resources), and so on. The external problems are more a combination of the Soviet legacy, which has conditioned the attitudes of governments and the population at large to question the relevance and priority of science, and current economic and political circumstances, which have led to cutbacks in governmental funding and support and a further questioning of the priority of science and technology in the light of other problems facing the countries.

These problems notwithstanding, a second significant conclusion to emerge from this volume is that the system has much to recommend it, both in terms of the real quality that it brings with it from the previous era and some very real progress which has already been made in identifying and implementing needed reforms. Despite a certain level of inertia or outright resistance to change in certain quarters, all of the countries examined here evidence, on balance, a recognition of the need for change and efforts to move forward in this respect. Even without this conclusion, the arguments advanced in the introductory paper regarding the strategic importance for economic development of a viable science and technology sector supports the view that the sector is worthy of attention and that efforts must be made to preserve it in some form appropriate to current needs. This point is strengthened by the determination that a base to build on and a will to proceed already exist.

Challenges

In general, the papers here have focused on three categories of challenges that science and technology and higher education in Central and Eastern Europe face:

1. System challenges: These include the need for a new legislative framework; the establishment of a system of governance for institutions that includes defining the role of the government in sector management; the need to set system priorities; the need to forge private sector/industry links; the need to foster cooperation and mechanisms for cooperation; and the need to rationalize the size and composition of the sector.

2. Internal organization/content challenges: These include ques-

tions of internal governance; the integration of research and teaching; curriculum reform; staff rationalization; staff retraining and upgrading; alternative sources of funding; tuition fees; and degrees offered.

3.   Economic challenges: These include the issue of appropriate levels of governmental funding; the immediate pressures of cutbacks in funding; the loss of personnel due to falling real wages or low nominal wages; lack of funds for equipment, libraries, and so forth; rising costs due to inflation; and the implications of the generally depressed economic climate in combination with the host of competing claims on economic resources by all sectors and constituencies.

As the papers point out, on average most progress to date has probably been made on the fronts of legislation and management systems, although those require further attention in most cases. But the emphasis, not surprisingly, has been more on process than substance, and the latter will be far more difficult to address or solve.

From a different perspective, the problems of science and technology and higher education can be characterized, in part, as a conflict between long-term needs and short-term constraints. This is especially true for the economic problems confronting the sector.

In the short term, the general economic climate is bad, and, in some cases, it threatens to grow worse. There are, moreover, a multitude of adjustment problems to be addressed, both at the macroeconomic and the microeconomic level, as well as a host of social issues that emerge directly from the economic adjustment process and that have immediate consequences for large numbers of people. These problems are obviously both political and social priorities. In this general environment of a declining economy and significant competing needs, the ability to do more for the science and technology sector is clearly constrained, quite apart from any question of the will to do so. At a more micro level, the same conflict between the short run and long run is present. In the short term, the system is constrained by existing personnel, equipment, institutional specializations, and so on. It is further constrained by attitudes and perceptions and strong vested interests. All of these factors are working to slow the pace of change or to steer it in inappropriate directions. Until these obstacles are overcome, genuine reform may be impossible.

In the long run, however, a strong, viable science-and-technology sector is important to economic growth and development, and therein lies the conflict between the short term and the long term at the macro level. Pursued to its logical conclusion, the short-term exigencies could well lead to the sector's being so severely weakened as to be made ineffectual for long-term purposes. Alternatively, full-scale attention to the long-term argument at the expense of achieving a more stable macroeconomic and political environment in the short term could leave the sector with little upon which to impact in the long run.

Lessons

What, then, are the lessons to be learned and the directions to be taken?

First, the process of change must be maintained and strengthened, but this must be done with the full recognition that the general economic and political environment is important to the success of this effort and cannot be ignored; thus expectations as to the pace of reform must be realistic.

Second, there is much for institutions and countries to learn from one another. As the papers in this volume well illustrate, each country has some lessons which, if learned, may help others. One of the significant problems of all these countries at the onset of the reform process was the relative isolation of the Soviet bloc from the rest of the world. In combination with that was the homogeneous quality of the Soviet system, which meant that countries lacked much meaningful comparative experience on which to draw in shaping their reforms. Drawing on one another's experience, as well as learning from the rest of the world, may be of tremendous value at this point.

Third is cooperation:

- Between institutions in each country;
- Between institutions of different countries of the CEE and FSU;
- Between institutions in the CEE/FSU and Western countries;
- Between institutions and the private sector;
- Between institutions and governments.

Each of these types of cooperation has a role to play in the reform process or will be crucial to the success of science and technology and higher-education reform. The simple message is that no one institution or country can succeed alone.

Fourth, government has a crucial role to play in the short run, both in creating much of the legal, economic, and political framework necessary for the reform process to proceed and in providing funding to maintain and upgrade the system as the reforms unfold. This does not mean maintaining the status quo but rather recognizing that, without appropriate funding, reform will falter both because many reforms will cost money, at least in the short run, and because appropriate economic incentives may be one of the surest ways of encouraging reform. Governments also have a crucial role to play in helping to determine the shape of the new system, but they should do so in cooperation with the institutions and personnel involved rather than attempting to impose solutions unilaterally.

Fifth, government will have an ongoing role to play in the support of, and in helping to manage, the science and technology sector, including the setting of science and technology policy. In other words, a pure market model will not work for the science and technology sector and should not be adopted. While the specific level and form of governmental involvement is open to debate, there is a minimal role that government must play in the long term if the full social and economic benefits of the science and technology sector are to be realized.

Most of these points are made not only in various of the country papers in this volume but also in the comparative examination of World Bank experience described in the Eisemon paper. That paper ends with a call for World Bank (and other donor) investment support for science and technology and higher education in CEE/FSU, recognizing that "even under the most favourable circumstances, multiple investments will be required to support the reform process in Eastern Europe." Eisemon goes on to argue the need for each government to formulate (perhaps with World Bank help) a comprehensive investment strategy for the sector.

The need for external support and the desirability of a planning exercise that defines needs and priorities are ideas that are also to be found, in some fashion, in several of the other papers. Together they compose the sixth and final of the lessons or future directions listed here.

This point has been left to last purposely, because it is in many ways the most important. It is, on the one hand, an appeal for help that is far from trivial in the context. But more significantly, it is a way of underlining two very important points that emerge from the various papers in this book:

1.   Countries need to find their own solutions, because, however else they are reached, they will not be accepted and will not succeed if they are not "locally owned"; this will require a process of planning and consensus-building within institutions, between institutions, with government, and so on.
2.   While solutions must be local in the sense just described, external assistance will be required both in terms of the planning exercise itself and in terms of plan implementation. Such assistance will be financial in many cases, but it may also come in other forms, such as advice, training, materials, or exchanges.

One of the important areas where outside help could be particularly useful at the present time is in helping to carry out the large amount of research and study necessary for a proper determination and assessment of the status quo and the goals, objectives, and means for change. The kind of research required is perhaps more akin to action research and evaluation than that which deals with policy options.

Final Comment

We end where we started, with the assertion that science and technology do matter to the larger process of reform and recovery now underway in Eastern and Central Europe and the Former Soviet Union. They matter not only in the sense of having a positive contribution to make but also in the sense of posing negative consequences if ignored or treated inappropriately. Recognizing this role and harnessing it for the good of the country is one of the greatest challenges the sector faces. Whether, and how, the challenge is successfully met will depend on many factors. The sector itself—that is, the institutions and personnel of higher education and science and technology—must play a major role in defining its future shape and direction, together with governments, students, the general population, and foreign experts and donors. All stakeholders must be involved and must recognize the legitimacy of each other's involvement.

The ingredients for a successful transition can thus be identified as consensual recognition of the importance of the sector; participation of, and cooperation among, all stakeholders; and external support. With these ingredients in place, the challenges can be met.

Index